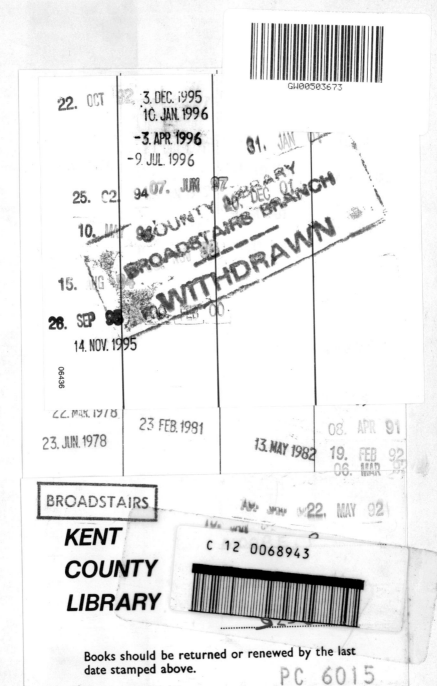

GUIDE TO THE MOON

GUIDE TO THE MOON

by

Patrick Moore, O.B.E., D.SC.(HON.), F.R.A.S.

Director of the Lunar Section of the
British Astronomical Association

LUTTERWORTH PRESS
GUILDFORD AND LONDON

This completely revised and new edition first published 1976

To my Mother

—since without her, during the last 52 years, neither this book nor any of my others would have been written!

ISBN 0 7188 1720 6

*Printed in Great Britain
by Ebenezer Baylis and Son Ltd.
The Trinity Press, Worcester, and London*

CONTENTS

		PAGE
Foreword		9
Acknowledgements		10
1	INTRODUCTION: 'The *Eagle* Has Landed'	11
2	THE LUNAR WORLD	14
3	A PICTURE OF THE UNIVERSE	21
4	THE BIRTH OF THE MOON	31
5	THE MOVEMENTS OF THE MOON	45
6	THE MOON AND THE EARTH	60
7	OBSERVERS OF THE MOON	74
8	FEATURES OF THE MOON	96
9	THE CRATERS OF THE MOON	110
10	ATMOSPHERE AND LIFE?	120
11	ECLIPSES OF THE MOON	135
12	THE WAY TO THE MOON	144
13	EXPLORING THE MOON	161
14	THE MOULDING OF THE SURFACE	173
15	FLASHES, GLOWS AND MOONQUAKES	197
16	BEYOND APOLLO	213
	Appendices	
I	Observing the Moon	217
II	Lunar Literature and Lunar Maps	225
III	Numerical Data	227
IV	Eclipses of the Moon, 1976–1986	228
V	Lunar Landings	230

CONTENTS

VI Description of the Surface, and Map   232

    First Quadrant   234
    Second Quadrant   251
    Third Quadrant   266
    Fourth Quadrant   285

VII Index to Formations in the Map and Text   306
    Latin and English Names of the
      Lunar Seas   312

VIII The Far Side of the Moon   313

    Index   316

LIST OF PLATES

(Plates I to XVI between pages 160 and 161)

I. BASALTIC LAVA-FLOWS

II. TYCHO

III. LUNAR FEATURES
(top left) Alphonsus and Arzachel
(top right) Mare Serenitatis
(below) Messier twins

IV. CRATERS ON MARS, MERCURY AND THE MOON
(top) The surface of Mars
(lower left) Craters on Mercury
(lower right) Mare Vaporum area of the Moon

V. LUNAR WALLED PLAINS
(top left) The ruined plain Janssen
(top right) Archimedes, Aristillus and Autolycus
(lower) Schickard and Phocylides

VI. THE LUNAR SURFACE: APOLLO AND LUNA
(top) Apollo 12 and astronaut
(below) Panoramic view from Luna 13

VII. EARTHSHINE AND LUNAR ECLIPSE
(top) Moon at partial eclipse
(below) Moon at total eclipse

VIII. THE FAR SIDE OF THE MOON
(top) The Moon from Lunik 13
(below) The crater Tsiolkovskii

IX. THE MOON FROM MARINER 10

X. THE SINUS IRIDUM

XI. THE MOON; ALMOST FULL

XII. ARISTARCHUS
(top) From Apollo 15
(below) From Earth

XIII. PTOLEMÆUS TO PLATO

XIV. THE MOON FROM APOLLO 16

XV. LUNAR RILLS
(top) Hyginus Rill
(below left) Schröter's Valley
(below right) Another Orbiter view

XVI. THE MOON FROM APOLLO
(top) From Apollo 14
(below) Apollo 17 landing site

FOREWORD

THE FIRST EDITION of *Guide to the Moon* was written soon after the end of the war, when the Moon was still very much of a Cinderella in astronomical circles; few professional scientists paid much attention to its surface, and space-flight was still regarded as a wild dream. Later editions of the book reflected a steady change of attitude. The last of these came out after the start of the rocket age, under the title of *Survey of the Moon*; Luna 3 had made its circum-lunar flight, but manned travel was still in the planning stage.

Since then Apollo has come and gone, bringing a complete revolution in our ideas about the Moon. I have therefore re-written the book completely, and reverted to the original title —though practically nothing of the first edition text is left!

As an observer, I have naturally written from the observer's viewpoint, and I have also tried to put matters in the proper perspective by relating some of the events which have led up to the present-day situation. Because I have my own ideas about various controversial topics, I have made no attempt to be unprejudiced—and I am very ready to be proved wrong. No doubt things will be as different by A.D. 2000 as they now are from the time when I began studying the Moon forty years ago.

PATRICK MOORE

Selsey, *April 1976*

ACKNOWLEDGEMENTS

MY SINCERE THANKS are due to the many people who have helped in the compilation of this book. I am most grateful to Commander Henry Hatfield, R.N., and Peter Foley for making their splendid lunar photographs available, and to NASA for the use of the Apollo and Orbiter photographs. Lawrence Clarke has not only provided all the line drawings, but has also completely re-drawn the lunar maps, which was a tremendous task—carried out with his usual skill. John Bunn, of Lutterworth Press, has been of great help in many ways. Last, but by no means least, I must thank Michael Foxell— without whom the book would never have appeared, and to whom I am deeply grateful.

PATRICK MOORE

Selsey, *January 1976*

Chapter One

INTRODUCTION: 'THE *EAGLE* HAS LANDED'

ON THE EVENING of 20 July, 1969, I was in a B.B.C. television studio. It was a great occasion: Astronauts Neil Armstrong and Edwin Aldrin were on their way to the Moon, and for the first time in human history the gap between the two worlds was about to be bridged.

I had the feeling that I was 'sitting in' at one of the great moments in the story of mankind—as indeed I was. I remember, too, recalling what a reviewer had said about a comment of mine made in the first edition of *Guide to the Moon*, published a quarter of a century ago, when I had suggested that the first lunar voyage would be made well before 1990. 'Even though the author's ideas about space-travel appear somewhat naïve . . .'

But uppermost in all our minds, of course, was the well-being of the astronauts themselves—not forgetting Michael Collins, the third member of the Apollo 11 team, who was patiently orbiting the Moon in the command module of the space-craft. There had been strongly-supported suggestions that the lunar maria, or dry 'seas', were nothing more nor less than treacherous dust-bowls into which a landing vehicle would sink with devastating permanence. Automatic craft had more or less ruled out this prospect, in which I had never had any personal faith, but it was still possible that there could be dangerous areas, and if the *Eagle* module made a faulty touch-down there could be no hope of rescue. It was a tense moment. The astronauts were too busy to pause and reflect (as they have since told me!) but the tension in Mission Control, Houston, must have been unbearable. It was nerve-racking enough even in our studio at Lime Grove.

Then came the last moments—and Neil Armstrong's voice: 'The *Eagle* has landed.' The surge of relief spread from Houston to the whole of the world. I have no clear recollection of what I said; we were on the air 'live', so I hope that my comments were coherent. Man had reached the Moon at last, and the dream of so many centuries had come true.

To me, that was the supreme moment of the expedition. Neil Armstrong's later words as he walked on to the lunar surface—'That's one small step for a man, one giant leap for mankind'—will be remembered as long as *homo sapiens* last, but I am bound to admit that I found it something of an anti-climax to 'The *Eagle* has landed'. The other moment of extreme tension came when the two astronauts were ready to blast away from the Moon, to rejoin Collins in lunar orbit. They were entirely dependent upon the single ascent engine of the lunar module, and if this engine had failed—well, the results would have been too appalling to contemplate. Mercifully, it worked without a fault.

Since then there have been six more manned expeditions to the Moon, of which five have been successful. There have been no casualties during the American space-programme, apart from the disastrous fire at Cape Canaveral (then Cape Kennedy) which took place during a ground rehearsal. The Russians have lost four men. Therefore, the fatal accidents during the space-programme to date have been far less than those during the development period of aeronautics. Admittedly there is no real comparison; aeroplanes were built by amateurs working on their own, while space-craft development needs the full resources of a Government. Yet the fact remains that so far we have escaped lightly, particularly since space is so hostile an environment.

There is no longer anything futuristic in talking about a Lunar Base, set up on the Moon's surface and manned by a permanent staff. During a *Sky at Night* television broadcast in 1970, Neil Armstrong gave his views without hesitation: 'I'm quite certain we'll have such bases in our lifetime.' We have come a long way during the past couple of decades. Remember, the Space Age began as recently as 4 October, 1957, when the Russians launched their football-sized Sputnik 1 and ushered in a complete revolution in human outlook.

When I began studying the Moon, in the 1930s, there was no official thought of anything of the kind. Space-travel was regarded as an amiably eccentric idea, and Interplanetary Societies were classed in the same category as the Flat Earthers (flying saucers, I may add, lay in the future). Professional astronomers paid scant attention to the lunar surface, and

Moon-mapping was largely the province of amateurs. Charts were drawn up; photographs were taken, and we did our best to find out the sort of world that the Moon might be. We were, of course, limited to studying less than 60 per cent. of it, because the remainder is always turned away from us, and until the flight of the Russian probe Luna 3 in 1959 we knew nothing about it.

The situation altered during the 1950s, when it became clear that the Moon was not, after all, completely out of reach. And today things are as different as they could be; Earth-based studies of the Moon are naturally limited in use, because we can study samples of lunar rock in our laboratories. Many major problems have been solved. Most important of all, perhaps, we can now be quite sure that there has never been any life on the Moon.

I am essentially a lunar observer, and I make no pretence of being knowledgeable about rocketry. To use scientific slang, I am no 'hardware man'. In this book, then, I will try to present a concise account of the Moon, tracing the story of how it has been studied, giving some of the old theories and fancies, and leading on to the present day. Much will necessarily be left out, and in the areas of my own research I will make no serious attempt to be unprejudiced. But there is a great deal to say; and so for a start let us go back to the remote past, when the very nature of the Moon was still unknown, and our remote ancestors were even disposed to regard it as the abode of gods.

Chapter Two

THE LUNAR WORLD

THOUSANDS OF YEARS AGO, before the start of recorded history, Stone Age men must have looked at the Moon and wondered just what it was. It was far larger and brighter than any of the stars; it moved quickly across the sky, changing shape regularly from a slender crescent to a full disk and then back again; it took second place only to the Sun. Surely it must be a god, or at least the home of a god?

Moreover, ancient peoples found the Moon very useful. In those far-off times, when lack of alertness was usually fatal, dark nights were the most dangerous ones, and the Moon's radiance gave some defence against surprise attacks by animal or human enemies. It is not in the least surprising that Moon-worship held an important place in primitive religion.

The Moon was also helpful in the measurement of time. The interval between one full moon and the next was found to be constant, as any casual observer will know, and the first rough calendars were drawn up to conform with it. At a very early stage it was also found that the ocean tides are regulated by the Moon, although the reason was unknown (and remained unknown until the time of Sir Isaac Newton).

Moon-myths probably go back almost as far as Man himself. Some of the old stories are fascinating, and every country seems to have its own legends. For instance, who has not heard of the Man in the Moon? It is true that by using one's imagination the dark patches on the lunar disk can be twisted into something like a human form, and the various myths are remarkably alike. According to a German tale, the Old Man was a villager caught in the act of stealing cabbages, and placed in the Moon as a warning to others. Another version, from the island of Sylt, makes him a sheep-stealer, and he is also a thief in a legend from Polynesia in the South Seas. Then, too, there is the story that the Old Man committed the sin of gathering faggots on a Sunday, and was given the choice between burning in the Sun or freezing on the Moon; after due consideration, he opted for the latter. The Chinese had a rather different idea. To them, the Old Man was to all intents and purposes the god of marriages, and it was his solemn duty to link husband and wife with an invisible silken thread which could not be broken until one of the partners died.

Women, of course, could not be left out. I particularly like another of the Polynesian legends, according to which a girl named Sina was unwise enough to compare the Moon with a huge bread-fruit—with the result that the infuriated Moon reached down and snatched up both Sina and her child, so that they can still be seen in the Moon today.

Animals also found their way to the Moon, not always of their own free will. From India comes the remarkable tale of how a wolf fell violently in love with a toad, and refused to take 'no' for an answer; to escape, the toad jumped on to the Moon and stayed there. (The wolf's comments are not on record.) Also from India we hear how a hare offered to cook himself in order to feed a hungry Brahmin, and was placed in the Moon as a reward for his courtesy.

To the people of Van, in Turkey, the Moon was a young bachelor, and was engaged to the Sun. Originally the Moon had shone in the daytime and the Sun at night; but the Sun, being a girl, was afraid of the dark, and persuaded her fiancé to change places. In another Turkish tale, the Moon was very fond of his mother, and used to follow her about everywhere, greatly to her annoyance. Once he followed her when she was washing dishes, and the mother was so angry that she threw the dish-cloth in his face, which explains why the Moon's disk now appears stained.

Even Greenland has its lunar myth. This time the Sun and Moon were brother and sister, and for some reason or other the Sun rubbed soot in his sister's face. Rather naturally, the Moon chased him—but she can never catch him, because she cannot fly so high. Every few weeks she feels the need for a rest, so she comes back to the ground, climbs into a sleigh drawn by four dogs, and goes seal-hunting. After eating several seals she regains her strength, and the Moon once more becomes full, so that the chase can start all over again.

Yet another Chinese legend explains how there was once a great drought, and a herd of elephants came to drink at a sheet of water called the Moon Lake. They trampled down so many of the local hare population that when they next appeared, a far-sighted hare pointed out that they were annoying the Moon-goddess by disturbing her reflection in the water. The elephants agreed that this was most unwise, and departed hastily.

True Moon-worship still lingers on in parts of Central Africa, and originally the Moon was regarded as one of the most powerful of all the gods. Generally—though not always—it was male, and only the Sun was more important. The ruins of a large lunar temple have been uncovered at Ur, while the Egyptians had two moon-gods, Khonsu (also the God of Time —here we have a link with the calendar) and Thoth. In Greece, Artemis was the lunar deity, while the Japanese moon-goddess rejoiced in the name of Tsuki-yomo-no-kami. The Aleutian Islanders were in the habit of stoning to death anyone who had been incautious enough to offend the Moon. And from the Confessional of Ecgbert, Archbishop of York, we learn that the British Druids still paid homage to the Moon as late as the eighth century A.D.

Quite apart from these myths, the early peoples managed to find out at least something about the Moon itself. At first they believed that it actually changed shape from night to night—in Bushman mythology the Moon was believed to have offended the Sun, and is regularly pierced by the Sun's rays until he pleads for mercy and is gradually restored!—but it was then found that this could not be so, because the 'dark' part of the disk can often be seen shining faintly alongside the brilliant crescent. In fact, this effect can be seen almost every night when the Moon is crescent-shaped; some people call it 'the Old Moon in the Young Moon's arms'. As Leonardo da Vinci pointed out later, it is due to light reflected on to the Moon from the Earth. At least it showed that the Moon is always a complete disk.

On the other hand, the ancients had no idea of the plan of the universe, and some of their ideas about the Earth itself were decidedly peculiar. Usually the world was a flat plate, some-times floating in water and sometimes surrounded by a solid sky. According to the Hindus it stood on the back of four ele-phants, which in turn rested upon the shell of a vast tortoise swimming in a boundless ocean. One cannot help feeling some-what sorry for the tortoise, but there were other theories too.

One Indian cult taught that the Earth is supported on twelve massive pillars, so that during the hours of darkness the Sun has to thread its way through a kind of maze without touching any of the pillars.

In Egypt, the priests who ranked as the country's leading scholars made the initial mistake of supposing that the universe takes the form of a rectangular box, with the longer sides running north–south. There is a flat ceiling, supported by four pillars which are connected by a mountain chain; below this chain lies a ledge containing the celestial river Ur-nes, along which sail the boats carrying the Sun, Moon and other gods. When a boat comes to one of the sharp corners of the river, it turns abruptly at right angles and continues blithely upon its way.

All this is intriguing, but it is not science; and true science did not begin until the time of the Greeks. Neither was it quick to develop. The first great Greek philosopher, Thales, was born shortly before 600 B.C., while Ptolemy, the last famous scientist of Classical times, died about A.D. 180. This gives us a total time-span of 800 years, so that chronologically Ptolemy was as far away from Thales as we are from the Crusades.

Thales believed the world to be shaped like a log or cork, and to be floating on water. His younger contemporary, Anaximander, had definite ideas about the Moon, and described it as 'a circle nineteen times as large as the Earth; it is shaped like a chariot-wheel, the rim of which is hollow and full of fire, as that of the Sun also is; it has one vent, like the nozzle of a pair of bellows; its eclipses depend upon the turnings of the wheel.' On the credit side, Anaximander did at least claim that the Earth is suspended freely in space without being held up by pillars, elephants, tortoises or anything else.

It would be too much of a digression to describe many of the other ideas current during the first centuries of Greek greatness, but I cannot resist quoting Xenophanes, who died in or about 478 B.C. at the advanced age of nearly one hundred. 'There are many suns and moons according to the regions, divisions and zones of the Earth . . . the Earth is flat. On its upper side it touches the air; on the underside it extends without limit.' Xenophanes believed that the various suns, moons and stars were made of clouds set on fire by some process which he did not describe. At about the same time another philosopher, Heraclitus, wrote that the diameter of the Sun is about twelve inches, which is something of an underestimate!

Gradually, the idea of the Moon as a body moving round the spherical Earth at a relatively slight distance began to gain support. It was also found that it has no light of its own, and depends upon light reflected from the Sun. Around 270 B.C. Eratosthenes of Cyrene, librarian in charge of the great collection of books at Alexandria, made a remarkably accurate measurement of the size of the Earth, and his contemporary Aristarchus had a very shrewd idea of the Moon's distance from us. (Aristarchus was also one of the first to suggest that the Earth moves round the Sun, but at the time he found few followers.) Lunar eclipses were correctly explained as being due to the Moon's entry into the shadow cast by the Earth, and it was taught that the markings on the disk were due to lofty mountains and deep valleys.

What about the chances of life on the Moon? To some of the Greeks there seemed every reason to suppose that beings of some kind lived there, though whether these beings were human or merely 'spirits' was an open question. Around A.D. 80 the celebrated writer Plutarch produced *De Face in Orbe Lunæ* (On the Face in the Orb of the Moon) in which he maintained that the lunar world is 'earthy', with mountains and ravines; but he was quite convinced that it must be inhabited. Space-travel ideas also go back to the Greeks, though the suggested methods were not of the kind calculated to appeal to NASA or to Neil Armstrong. I will have more to say about them later.

The last great Classical philosopher, Ptolemy, wrote a book in which he summed up the astronomical knowledge of his time. Because his text has come down to us only via its Arab translation, we usually know it by its Arab title: the *Almagest*. In it, Ptolemy described how the celestial bodies move round the Earth in a rather complicated manner. Like virtually all his contemporaries he believed that all paths or orbits must be perfectly circular, but he was too good an observer and mathematician to think that the situation was straightforward, and he had to introduce various refinements which need not concern us at the moment. His system—always called the

Ptolemaic, though Ptolemy himself did not invent it—was generally accepted for more than a thousand years after his death, but unfortunately it was completely wrong. Astronomy could make real strides only when the idea of a motionless, central Earth had been cleared out of the way.

Actually, the Moon is the only natural body which really does move round the Earth—and even this is an over-simplification. The planets revolve round the Sun, while the stars are suns in their own right. Few of the Greeks had the courage to dethrone the Earth from its proud central position (Aristarchus was an exception), and the great upheaval in human thought was delayed until less than five hundred years ago, which is a sobering thought. I am trying to do no more than give a brief sketch of events, so we can safely skip more than a dozen centuries and come to what is generally called the Copernican Revolution.

It began in 1543, with the appearance of a book, *De Revolutionibus Orbium Cœlestium* ('Concerning the Revolutions of the Celestial Orbs'). The author—Copernicus, a Polish canon —prudently held back publication until he was dying, because he was well aware that the Christian Church would not take kindly to the idea of a moving Earth; and in this he was correct. The 'Copernicans' were savagely persecuted, and one of them, Giordano Bruno, was burned at the stake in Rome in 1600. What really tipped the scales was the work of a German mathematician, Johannes Kepler, who spent many years in studying observations of the planet Mars made by Tycho Brahe, an eccentric Dane who was firmly committed to the idea of a central Earth but whose measurements of the positions of the stars and planets were amazingly accurate. Kepler found that the planets move round the Sun not in circles, but in ellipses. His famous *Laws of Planetary Motion*, published between 1609 and 1618, gave the death-blow to Ptolemy's theory. It was during this period, too, that telescopes were turned toward the sky; for the first time men could see that instead of being covered with grassy plains, glittering oceans and spreading forests, the Moon was a world of rugged mountains, broad plains and huge craters.

Apparently the first man to look seriously at the Moon through a telescope was Thomas Harriott, one-time tutor to

Sir Walter Raleigh, but the real pioneer was Galileo, who described the lunar surface in great detail and even made some reasonably successful attempts to measure the heights of the mountains. Galileo's work may be said to have ushered in the 'modern' era, though the old ideas were far from dead. Despite the detailed maps which were drawn up during the following years we still find Sir William Herschel, one of the greatest of all observers, maintaining that the Moon was certainly inhabited—and Herschel died as recently as 1822. But as time went by, and the airless, hostile nature of the Moon became painfully obvious, opinions changed. Well before the first rockets were sent there, the idea of lunar creatures had been firmly relegated to the scientific scrap-heap.

So far as space-travel was concerned the Moon had to be the first target, because it is so near at hand and because it keeps company with us in our never-ending journey round the Sun. The Apollo astronauts showed the way, and signals are still being received from the scientific equipment that they left behind. We can look ahead to the Lunar Base with modest confidence. But before taking the story any further, it seems only right to pause momentarily and try to put the Moon in its proper perspective with regard to the Earth and the other bodies in the universe.

Chapter Three

A PICTURE OF THE UNIVERSE

THE MOON IS a splendid object in our skies, and we naturally tend to think of it as important. Obviously it cannot compare with the Sun, and the difference is greater than most people realize. Sunlight is over half a million times more powerful than the radiance of the full moon, and it is quite wrong to think that a moonlit night can be almost as bright as day. Yet the Moon is both beautiful and imposing, and means much more to us than the tiny, twinkling stars.

Modern astronomy paints a very different picture. Things are emphatically not what they seem, and of all the celestial bodies visible with the naked eye the Moon is the least important (unless we count cosmical débris and our own artificial satellites). Officially the Moon is not ranked as a planet, and is said to be a mere junior attendant of the Earth (Fig. 1). I am not at all sure that this view is justified, for reasons I propose to give below, but it is true that even Mercury, the smallest of the planets, is larger and more massive than the Moon.

What makes the Moon unique is its closeness to us, and the fact that it stays near us all the time. Its average distance from the Earth is rather less than a quarter of a million miles, and the nearest of our planetary neighbours—Venus, incidentally, not Mars—is always at least a hundred times as remote. The distance of the Sun is about four hundred times that of the Moon. And when we come to consider the stars, we are faced with distances so great that we can have no hope of appreciating them.

Can you really imagine what is meant by 'a million miles'? I admit that I cannot, and a million miles is a very short stretch on the scale of the universe. Astronomers cannot appreciate great spans of space and time any better than laymen—the only difference is that they do not make the mistake of trying! Instead of using actual figures, it may be of some help to give a scale model, starting with the Solar System in which we live.

The Solar System is made up of one star (the Sun), nine

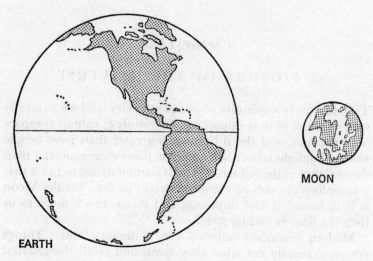

MOON

EARTH

Fig. 1. Comparative sizes of Earth and Moon

known planets, and various lesser bodies, including the satellites or moons of some of the planets. Only the Sun has light of its own, so that the other members of the system have to make do by reflecting the solar rays, in the manner of large but not very efficient mirrors. (The Moon is particularly poor as a reflector of light; in scientific parlance, it has a very low albedo of only about 7 per cent.) Let us begin by making the Sun a globe two feet in diameter, and putting in the planets to scale.

The innermost member of the family, Mercury, will become a grain of mustard seed, moving round our model Sun at a distance of 83 feet. Venus is represented by a pea at 156 feet, and the Earth by another pea at 215 feet. Mars, the outermost of the small planets, becomes a small bead at 328 feet. Before going any further, what about the natural satellites? Mercury and Venus have none; our Moon becomes another seed, moving round the Earth at a distance of 6½ inches; and Phobos and Deimos, the two dwarf attendants of Mars, will be so tiny on this scale that a microscope would be needed to show them at all.

All the four inner planets are solid and rocky, and have various points in common. Mercury is not a great deal larger than the Moon, and has proved to be amazingly 'lunar' in

22

type. Photographs of it taken in 1974 and 1975 from the Mariner 10 space-craft have shown it to be covered with mountains and craters; look at one of the Mariner pictures, and you will find it hard to tell whether you are seeing Mercury or the Moon. Venus is almost the Earth's twin in size and mass, but it is a very odd world indeed. It has a dense atmosphere made up chiefly of the heavy gas carbon dioxide, together with droplets of uninviting substances such as sulphuric acid, while the surface temperature is not far short of 1,000 degrees Fahrenheit. Craters have been detected by radar and in 1975 rocks on the surface were photographed from Russian soft-landing probes. Mars too has its craters, well shown on the pictures sent back from the various space-craft which have been close-in. The best results so far come from Mariner 9, which was launched in 1971 and put into a path round Mars; for months after arrival it sent back an impressive series of photographs, and we now know a great deal about the wide craters, the deep valleys and the towering Martian volcanoes. Mariner 9 also paid due attention to the two satellites, finding craters there as well.

In some ways Mars has been a disappointment. The famed canal system does not exist, and the dark patches are certainly due to mere differences in surface colour rather than to vegetation, so that the chances of our finding advanced life-forms are virtually nil. Neither could we survive there unprotected, because Mars is a chilly world, and the carbon-dioxide atmosphere is much more rarefied than the air on top of Mount Everest. Yet the Red Planet is geologically fascinating, and it will almost certainly be a target for astronauts during the twenty-first century.

So far as the Moon is concerned, the most interesting point is that all these worlds—Mercury, Venus, Mars, Phobos, Deimos —are cratered. It is the Earth which is exceptional, inasmuch as our own craters are relatively few and small. There are excellent reasons for this, but for the moment let us press on to those regions of the Solar System beyond the path of Mars.

Going back to our scale model, we come next to the asteroids or minor planets, moving in paths about 1,000 feet from our two-foot Sun, and represented by grains of fine sand. Even the largest of them, Ceres, is no more than 700 or 800 miles in diameter (bigger than was thought until very recently), and most of the rest are lumps of material ranging down to less than a mile across. There are not many asteroids with diameters of over 150 miles, and only one member of the swarm, Vesta, is ever visible to the naked eye.

More than two thousand asteroids have had their orbits worked out, and the total membership of the swarm may be well over forty thousand. When the first space-probes to Jupiter were sent out, many people had uneasy feelings that crossing the rocky zone would be a distinctly hazardous business. In fact both Pioneer 10 and Pioneer 11 passed through unscathed, so that the danger may well be less than had been anticipated.

Some of the asteroids have unusual paths which swing them away from the main swarm. For instance there is Eros, a sausage-shaped worldlet less than 20 miles long, which came within 15,000,000 miles of the Earth in 1975, though there was never any fear that it would hit us. In 1937 an even smaller asteroid, Hermes, brushed past the Earth at a mere 485,000 miles, which is less than twice the distance of the Moon. Icarus, not much larger than Hermes, can actually go closer-in to the Sun than Mercury, while on the other hand the Trojan asteroids move at a mean distance equal to that of Jupiter.

Nobody is quite sure about the origin of the asteroids. They may be the remains of an old planet (or planets) which met with some disaster in the remote past, and broke up, though they may be merely the débris left over when the principal planets were formed. In any case they are insignificant, and lumping them all together would not make one body as massive as the Moon.

Next in order come the four giants of the Solar System. In our scale model, Jupiter becomes an orange one-fifth of a mile from the central Sun; Saturn, a tangerine at two-fifths of a mile; Uranus, a plum at four-fifths of a mile; and Neptune, another plum at $1\frac{1}{4}$ miles. All these great worlds are gaseous, and so there is no chance of landing on them. Of more interest to space-planners are the satellite families. Jupiter has thirteen attendants, Saturn ten, Uranus five and Neptune two, but not all of them are large.

In the Jovian family we have four satellites—Io, Europa,

Ganymede and Callisto—which are bright enough to be seen with any telescope; a few lynx-eyed people can see them with no optical help at all. Europa is slightly smaller than our Moon, Io slightly larger, and the other two much larger; indeed, both Ganymede and Callisto are comparable in size with the planet Mercury. The remaining nine satellites are tiny, and may well be captured asteroids rather than bona-fide satellites.

Of Saturn's attendants, much the largest is Titan, which is probably at least 3,000 miles in diameter, and whose atmosphere is much denser than that of Mars. All the satellites of Uranus are smaller than our Moon, but Triton, the senior satellite of Neptune, is distinctly larger.

Finally, on the boundary of the main Solar System, we come to Pluto, which was discovered as recently as 1930 and which is almost certainly smaller than the Earth. It has a path which is much less circular than that of any other planet, and on our scale it may wander out to more than two miles from the central globe. There are some strange problems connected with Pluto, and there are suggestions that it is nothing more than an ex-satellite of Neptune which broke free in some way—a kind of cosmical U.D.I.—and is now masquerading as a planet in its own right. It seems to be comparable in size with Triton, and must be intensely cold and lonely. If there is yet another planet, moving at a greater distance from the Sun, it must be so faint that its discovery will be largely a matter of luck.

So much for our scale model; it may help to add a few real values. The Earth is 93 million miles from the Sun; to either side of its orbit in the Solar System we have Venus (67 million miles from the Sun) and Mars (rather over 141 million miles). At its closest, Venus may approach us within 25 million miles, Mars within 35 million and Mercury within 49 million. These distances seem formidable when we remember that the Moon moves at a mere quarter of a million miles from the Earth.

In addition to the planets, the Solar System contains various bodies of lesser importance. For instance there are the comets, which are not solid and rocky, but are made up of small pieces of material together with thin gas. A bright comet may have a long tail (or tails), and in the past we have had some brilliant visitors, which used to cause a great deal of alarm and despondency inasmuch as they were regarded as evil omens.

Unfortunately all the really spectacular comets move round the Sun in highly elliptical paths, so that they have revolution periods amounting to many centuries, and we cannot predict them. Great comets have been rare during our own time, though we have had quite a number of comets bright enough to be seen with the naked eye. The most disappointing comet of recent years was Kohoutek's, which was expected to become striking around Christmas 1973, but which signally failed to do so. It may perhaps make a better showing on its next return—but we must resign ourselves to waiting for about 75,000 years.

There are plenty of faint comets which move round the Sun in short periods. Encke's Comet, for instance, is an old friend, since it comes back every 3·3 years, and we always know when and where to expect it. The only bright comet to have a period of less than a century is Halley's, which last came to perihelion (that is to say, its closest point to the Sun) in 1910, and which will be back once more in 1986.

Comets are flimsy things, and their solid parts seem to be made up chiefly of ice. Indeed, it has been said that a comet may be likened to a 'dirty snowball'. Because all comets move far beyond the Earth's air, they cannot be seen to shift perceptibly against the starry background, and if you see something crawling noticeably across the sky it cannot be a comet. It will probably be an artificial satellite.

Meteors are small particles revolving round the Sun in the manner of tiny planets. Most of them are smaller than pin's heads, and are too faint to be seen in the ordinary way; but if a meteor comes close to the Earth it dashes into the upper air, and becomes heated by friction, destroying itself in the streak of luminosity which we call a shooting-star, and ending its journey in the form of fine 'dust'. Meteors tend to travel round the Sun in shoals, and when the Earth passes through a shoal—as happens frequently—the result is a shower of shooting-stars. The August meteors, known as the Perseids, are particularly reliable and spectacular. There are also non-shower or sporadic meteors, which may appear from any direction at any moment.

Meteors are associated with comets, and can be classed as cometary débris. Meteorites, which are much larger, belong to a different category. They are too massive to be burned away as they plunge earthward, and they land as solid lumps, sometimes producing craters. Most museums have collections of meteorites which range in size from tiny specimens to large blocks, but occasional giants have also been found. The largest meteorite 'in captivity', so to speak, weighs 36 tons, and you can see it if you go to the Hayden Planetarium in New York; it was discovered in Greenland by the polar explorer Peary. The most celebrated meteorite to fall in Britain during recent years landed in the Leicestershire village of Barwell on Christmas Eve, 1965, and caused considerable excitement. It broke up during its descent, and fragments of it were scattered over a wide area. Before its disruption, the meteorite may have had a diameter of at least six feet.

Major falls are very rare, and we know of only two in the last few hundred years. The first was that of 1908, in the Tunguska region of Siberia, which blew pine-trees flat for miles all round the point of impact, though mercifully the area was uninhabited and there were no casualties (except in the local reindeer population). No meteoritic fragments have been found, and many of the Russian astronomers believe, with sound logic, that the missile was more likely to have been the head of a small comet than an ordinary meteorite. This would account for the lack of débris, because the icy materials making up a comet would be quickly evaporated, and would leave no trace behind. However, there is no doubt about the second fall, which took place in 1947 in the general region of Vladivostok; many meteoritic fragments were collected, and there were numerous small impact craters.

Other proven meteoritic falls of this magnitude are prehistoric. The holder of the heavyweight record is the Hoba West Meteorite, which is still lying where it fell near Grootfontein in Southern Africa; nobody is likely to run away with it, since it weighs at least 60 tons. You may also be interested to know that the Sacred Stone at Mecca is undoubtedly a meteorite, though we have no idea when it fell.

Go to Arizona, not too far from the town of Winslow, and you will find the famous Meteor Crater—strictly speaking, it should be Meteor*ite* Crater—which is almost a mile across, and of whose nature there is no doubt at all. The Wolf Creek Crater in Australia is also of impact origin, but other structures, such as the Chubb Crater in Quebec and the Vredefort Ring in South

Africa, are more problematical. I mention them here because many astronomers believe that the chief craters of the Moon were also produced by falling meteorites.

There is a fundamental difference between the shooting-star meteors and the true meteorites. On the whole it seems that there may be no genuine distinction between a large meteorite and a small asteroid, but there is no connection with comets. The materials are familiar enough; meteorites may be irons, stones, or a blend of the two.

Even apart from the meteoritic particles, the space between the planets is not empty. There is a great deal of thinly-spread material, lying chiefly near the main plane of the Solar System and causing the faint glows which we call the Zodiacal Light and the Gegenschein. Studies carried out from space-probes have given us considerable information about it.

So much for the Sun's family. It is our home, which makes it of vital importance to us, but it makes up only a very small part of the universe. Beyond Neptune and Pluto come vast stretches of space before we find the nearest stars; the absolute isolation of the Solar System is something which is hard to appreciate. For the last time let us go back to our scale model, and put our two-foot Sun in the middle of London. All the planets will lie well inside the Greater London area, but even the closest star will have to be represented by a globe somewhere in the frozen wastes of Siberia, and all but the half-dozen nearest stars will have to be placed clear of the Earth altogether.

We had better abandon our model, and look for a new unit of distance to replace the mile and the kilometre, which are much too short. Luckily there is a suitable unit available. As long ago as 1676 the Danish astronomer Ole Rømer found that light does not travel instantaneously; it races along at a velocity we now know to be 186,000 miles per second (or, if you want to be precise, 186,282·3959 miles per second—roughly 670 million m.p.h.). Therefore, light can leap from the Moon to the Earth in $1\frac{1}{4}$ seconds, and from the Sun to the Earth in 8·6 minutes, but the light from the nearest star beyond the Sun cannot reach us in less than $4\frac{1}{4}$ years. The distance covered by light in one year is rather less than six million million miles, and this unit, known as the light-year, is often used in measuring the distances of objects outside the Solar System.

A star is a sun—or, to put it more forcibly, our own blinding Sun turns out to be nothing more than a normal star, far less splendid than many of those we can see on any clear night. For instance, the brilliant white Rigel, in the foot of Orion, is at least 50,000 times as luminous as the Sun, though it is 900 light-years away, and we are now seeing it as it used to be in the time of William the Conqueror. On the other hand there are also stars which are much less powerful than the Sun. If we represent the Sun by an ordinary electric light bulb, the most luminous stars will be searchlights, while the feeblest will be tiny glow-worms.

All the stars are racing about at high speeds, but they are so far away that to all intents and purposes they seem to remain in the same patterns. True, their individual or 'proper' motions can be measured from year to year, but to the naked eye the constellations do not change even over hundreds of years; Orion, the Great Bear and the rest look the same now as they must have done to Julius Cæsar or even the builders of the Pyramids. Only the members of the Solar System are near enough to show obvious changes in position from night to night.

Our own system of stars, the Galaxy, takes the form of a vast spiral, 100,000 light-years from one side to the other; it contains about a hundred thousand million stars. Of course, it is not the only galaxy. There are plenty of others, most of them so remote that their light takes millions of years to reach us, and when we bear in mind that we can see only part of the universe it is very clear that the Solar System is very insignificant indeed. No doubt many of the other suns have planet-families of their own, and few modern astronomers doubt that life is widespread, though whether we will ever be able to contact it remains to be seen.

Certainly we cannot hope to get in touch with 'other men' by means of rockets. The Moon can be reached in a very few days, while it takes a modern space-craft less than a year to travel to Mars or Venus, and less than two years to range out as far as Jupiter. At the moment Pioneer 10, the first Jupiter probe, is on its way out of the Solar System altogether, but even before it passes the orbit of Neptune we will have lost all track of it, and not for many thousands of years will it be anywhere near another star. If we are to achieve interstellar travel it must be by some method about which our present ignorance is complete, so that speculation is both endless and pointless. I do not claim that it will never be done; I do say that it will remain impossible until we have made some dramatic 'breakthrough', which may not be for a long time yet.

I do not apologize for this somewhat lengthy digression, because it serves to show that we must avoid being parochial. The Solar System is unimportant in the context of the universe, and the Moon is a junior member of it. On the other hand it is of special interest to us, and within the next few decades we may have transformed it from a sterile into a living world.

Chapter Four

THE BIRTH OF THE MOON

EVERYONE IS INTERESTED in 'the beginnings of things', and we would very much like to know just how the Moon came into existence. As yet we cannot claim to any full understanding of the story, though it does seem that modern theorists are on the right track.

The whole question is linked with the origin of the Solar System itself. We have to start somewhere, and one thing we do know, with fair certainty, is the age of the Earth, which proves to be about 4,700 million years. There are various lines of investigation, all of which lead to much the same result. One of the most convincing methods of attack concerns the decay of certain very heavy substances, such as uranium. Uranium is radioactive, and decays gradually into lead, but it is in no hurry; its half-life—that is to say, the time taken for half the original uranium to change into lead—is over 4,000 million years. Fortunately, the lead produced by uranium decay is not quite the same as ordinary lead, and Nature has presented us with a very powerful research tool, because the quantity of uranium-lead associated with the remaining uranium tells us how long the process has been going on. (I realize that this is a gross over-simplification, but it will suffice for the moment.) The oldest Earth-rocks known to us have been found to be $4\frac{1}{2}$ thousand million years old, so that the world itself must date back further than this.

In 1796 the famous French astronomer Laplace proposed what was later called the Nebular Hypothesis, according to which the Solar System began its career as a vast, slowly-rotating gas-cloud. As it shrank under the influence of gravitation, various rings broke off, each of which condensed into a planet; the outer planets were therefore produced first, while the inner planets (Mars, Venus, Earth and Mercury) were younger. The Sun was regarded as the remnant of the original cloud, while the Moon was born from a gaseous ring thrown off by the contracting Earth.

31

Philosophers approved of the Nebular Hypothesis, but mathematicians had their doubts, and eventually they showed beyond all reasonable doubt that Laplace's theory, in its original form, simply would not work. Next—from 1900 onward—came the so-called tidal theories, according to which the planets were pulled off the Sun by the action of a passing star. These also were tested mathematically and were found to be wanting, plausible though they looked at first sight. Nowadays it is thought virtually certain that the planets built up by 'accretion' from a cloud of material which used to surround the Sun. There is a vague resemblance to the ancient Nebular Hypothesis, but the details are very different.

On the modern view, all the planets must be of roughly the same age, and they have evolved at different rates simply because they differ in mass. Jupiter, for instance, is over 300 times as massive as the Earth, and has a very strong gravitational pull, so that it has been able to hold on to all its original quota of hydrogen, the lightest of all the elements and the most plentiful substance in the universe. The Earth, on the other hand, has lost much of its hydrogen; and Mars, which has only one-tenth of our mass, has lost most of its original atmosphere as well. One often reads that Mars is 'a very old planet'. Undoubtedly it gives the impression of being senile, but its true age is probably about the same as ours.

Until recent times it was impossible to obtain definite proof, but when the first rocks were brought home from the Moon by Apollo astronauts the chemists and the geologists swarmed to the attack. The lunar samples proved to be between 3·3 and 4·5 thousand million years old, in excellent agreement with our results for the rocks of Earth. The two worlds must have been born at the same epoch, and there is no reason to doubt that when we manage to obtain samples from Mars, Mercury and other bodies we will find the same thing.

Armed with this information, we can start examining the various theories about the origin of the Moon itself. First there is the so-called fission hypothesis, originally proposed during the last century by G. H. Darwin, son of the great naturalist Charles Darwin. Few scientists nowadays have the slightest faith in it, but as it is still repeated in various popular books I can hardly ignore it.

Darwin started by assuming that the Earth and the Moon originally formed one body, and that the Moon was thrown off as a fluid-mass. In a modified version of this idea, the Earth had cooled down sufficiently to form a thin crust before the separation took place, and the sequence of events was worked out in considerable detail. The Earth, rotating rapidly on its axis, was in the state known as 'unstable equilibrium', so that it became egg-shaped, spinning about its shorter axis. Two main forces were acting upon it—the tides raised in it by the Sun, and its own natural period of vibration. When these two forces were in resonance (that is to say, acting together) the tides increased to such an extent that the whole body became first pear-shaped and then dumbbell-shaped, with one 'bell' (the Earth) much larger than the other (the future Moon). Eventually the neck of the dumbbell broke altogether and the Moon moved away, settling into a stable orbit (Fig. 2).

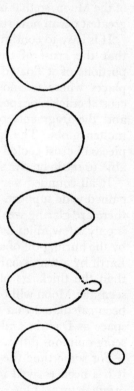

A strong supporter of the fission idea was W. H. Pickering, an American astronomer whose main interest was lunar work. Pickering, who died as recently as 1938, went even further than Darwin had done, and wrote that if the theory were correct the thin crust of the otherwise fluid Earth must have been torn apart, leaving a huge hollow where the thrown-off mass had once been. Moreover, the shock caused by the final fracture would have been violent enough to crack the fragile crust in other places as well.

This seemed to tie in well with the theory of continental drift, which has recently come into favour among geologists. A glance at any map or globe of the Earth shows that if the opposite sides of the Atlantic Ocean could be clapped together they would

Fig. 2. The Tidal Theory

fit into each other. Allowing for the sea having washed away portions of land here and there, and supposing Britain and the European mainland to be joined—as was the case only 10,000 years ago—the relationship is striking. The bulge of Africa fits into the hollow of South America, and the eastern coast of North America with the west coast of Europe. It seems that the continents may be regarded as 'rafts', floating around very slowly in the denser material below.

Pickering supported the continental drift theory, and was particularly interested in the Pacific Ocean, which, as he pointed out, is roughly circular. He believed that the great rounded hollow which now forms the bed of the Pacific is nothing more nor less than the scar left in the Earth's crust by the departure of the Moon, so that our satellite was born in the spot where our greatest ocean now rolls.

It is easy to continue the story. Pickering went on to explain that the crust of the Earth cracked under the shock, and portions of it floated apart, to settle down eventually in the places where we now find Eurasia and the Americas. The crustal cracking exposed the intensely hot interior of the Earth, and the fragments of crust floated as skin or scum on the molten globe. The lava surface exposed beneath the broken pieces of crust cooled and solidified, and later, when water was able to settle on the surface, became the Atlantic Ocean.

It all sounded very neat, and for a time Pickering's ideas gained wide support. Unfortunately, later work has shown that there are glaring weaknesses. To begin with, the Pacific Ocean is only a few miles deep, and can hardly represent the scar left by the hurling-off of a body the size of the Moon. Represent the Earth by a tennis-ball, and the depth of the Pacific will be less than the thickness of a postage stamp—whereas, as we have seen, the Moon will be the size of a table-tennis ball. It has also been calculated that a large mass could not be hurled off into space as Darwin and Pickering had supposed, though on this score opinions differ.

For some time the fission theory was regarded as dead, but it has been revived in recent years, though not in Pickering's original form. One variant involves Mars, whose diameter is just over 4,000 miles, intermediate between that of the Earth and the Moon. It has been suggested that Mars was thrown off the Earth and moved away independently, while the Moon is merely a droplet which was formed between the two bodies during the process of separation. But the main support for a fission theory has come from H. C. Urey and John O'Keefe, in America, whose ideas are based upon studies of the Moon's composition.

It is known that the Moon has a relatively small metallic core, no more than 450 miles in diameter at most insofar as the molten metal is concerned. The Earth's core, on the other hand, has a diameter of around 4,000 miles (Fig. 3). O'Keefe has commented that while the Earth and the Moon were joined as one body, the heavy iron content sunk to the centre, taking with it metals such as gold and platinum. The Moon was then formed from the Earth's outer layers (the mantle), which explains the smaller iron content and the lack of metals of the gold and platinum type.

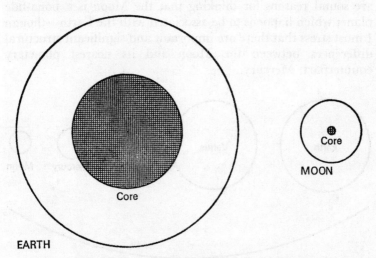

Fig. 3. Comparison in size between Moon and Earth cores

Results of this kind could not have been discussed before the Apollo lunar missions, because for our knowledge of the Moon's interior we depend upon the equipment left there by the astronauts—about which I will have much more to say later. There is no doubt that Urey and O'Keefe have made many interesting points. The fission theory cannot be dismissed;

but it is fair to say that on the majority view, the Earth and the Moon have always been separate bodies.

This brings me back to the possibility that the Moon is a companion-planet rather than a satellite. Admittedly there are some planetary satellites which are bigger than the Moon; three members of Jupiter's family, one of Saturn's and one of Neptune's. But all these satellites move round giant planets, and are very small compared with their primaries. Jupiter is over 80,000 miles in diameter, Saturn over 70,000 and Neptune over 30,000.

The Moon has one-quarter the diameter and 1/81 the mass of the Earth. Titan, the largest attendant of Saturn, has 1/20 the diameter and 1/4,150 the mass of its primary; for Triton in the Neptunian system the figures are 1/10 and 1/750 respectively. Moreover, the ratio Earth/Moon is not so very different from the ratio Venus/Mercury (Fig. 4), so that there are sound reasons for thinking that the Moon is a bona-fide planet which happens to be associated with the Earth—though I must stress that there are important and significant structural differences between the Moon and its nearest planetary counterpart, Mercury.

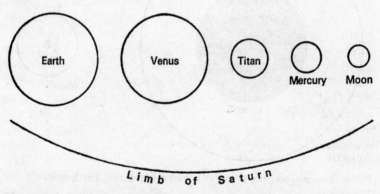

Fig. 4. Relative sizes of Earth, Venus, Titan, Mercury, Moon

On some theories the Moon was born from a cloud of material which once surrounded the Earth in a sort of ring, but it may well be that the Earth and the Moon came into existence at about the same time and in the same way; they are certainly of about the same age, as analyses of the Apollo rock samples show. The Moon may have been formed so close to the Earth that the two bodies have always been linked; alternatively it may once have been independent, and was captured by the Earth long after its formation. Here, too, opinions differ.

One point at least has been cleared up. Some authorities, including Professor Harold Urey, maintained that the Moon was never molten, and that the whole process of formation from the solar cloud was a 'cold' one; but since we now know that the core is reasonably hot, the opposing school of thought seems to have won the day. Of course, the temperature at the Moon's centre is lower than in the case of the Earth, but it is still appreciable. There is virtually no general magnetic field on the Moon now, though the evidence shows that there used to be one in the remote past.

Clearly we do not know nearly so much about the Moon's origin as we would like to do, and even after the Apollo missions we are still rather in the dark. Yet one thing is certain: the Earth-Moon system is unique in the Solar System.

There seems to be general agreement that at one stage, several thousands of millions of years ago, the Moon was closer to us than it is today. If so, then the 'month', or the time taken for the Moon to revolve round the Earth, must have been shorter; at a distance of about 12,000 miles the Moon would have completed one circuit in a mere 6½ hours. In this kind of situation, the tides raised by the two bodies upon each other would have been violent indeed. Even at its present distance the Moon still causes strong tides in our oceans, so that at 12,000 miles the effects would have been tremendous, but the Moon itself was the greater sufferer. These mutual tides had two important results. They slowed down the axial rotation periods of both the Earth and the Moon, and they pushed the Moon further away.

Initially, it is thought that the Earth's rotation period was much less than its present 24 hours, but this too was lengthened as the torn, tide-rent Moon receded. According to one set of calculations, the situation about 500 million years ago was that the Moon was already at least 200,000 miles away, and the persistent pull of the Earth had raised a permanent 'bulge' in the lunar globe, tending to keep this bulge turned Earthward and slowing down the Moon's rate of spin. The effect may be compared with that of an engine wheel rotating between two brake shoes (Fig. 5), though I admit that the analogy should not be taken too far. What happened was that the Moon, still not truly solid, had to fight against the tidal forces which were checking its rotation. The process went on and on, until at last the Moon had stopped rotating altogether relative to the Earth. By then its distance from us had grown to the present value; the revolution period was 27·3 days, and the time taken for the Moon to spin on its axis was precisely the same. Meantime, the Earth's axial rotation period had increased to almost 24 hours.

Perhaps I had better pause to say something about why these tidal effects tended to drive the Moon away. The crux of the

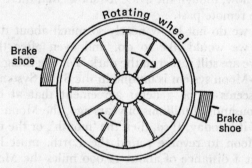

Fig. 5. Engine wheel between brake shoes

matter is what is called angular momentum. The angular momentum of a moving body is obtained by multiplying together the mass, the square of the distance from the centre of motion, and the rate of angular motion—that is to say, the rate of axial rotation. According to a well-known principle, angular momentum can never be destroyed; it can only be converted. If therefore the axial rotation slowed down, as happened in the Earth-Moon system because of tidal forces, something else had to increase, and this 'something' was the distance between the two bodies.

The process is not complete even now, because the Moon's tidal pull on the Earth is still braking our rotation. Each day is approximately 0·00000002 seconds longer than its predecessor, though there are also irregular fluctuations unconnected with the Moon. This works out to less than one second in a hundred thousand years, which is not very much, though the effects show up when we come to consider lunar eclipses observed many centuries ago. Also, the Moon is still receding from us. The rate of increase is only four inches per month, however, so that there is no hurry to study the Moon before it disappears into the distance!

Actually, however, this will never happen, because the Moon will not go on receding indefinitely. Various predictions have been made. For instance, it has been suggested that after moving out to more than 350,000 miles the Moon will start to come inward again, principally because of tidal effects due to the Sun, and will eventually be broken up by the Earth's gravitational pull, ending its career as a swarm of particles and providing us with a system of rings not unlike those of Saturn. Much has been heard of the Roche Limit, so called because it was first worked out by the French mathematician Édouard Roche. If a small body approaches its primary within a certain distance—less than about 10,000 miles in the case of the Earth —it may be disrupted. Strictly speaking, this applies only to a body which has no rigidity of its own, but it is certainly true that all the natural satellites of the planets lie outside the Roche limits for their primaries, whereas Saturn's rings lie inside.

Yet there seems no chance that the Moon will suffer so sad a fate. For one thing it is a rocky world, even though its density is less than that of the Earth, and we must not draw too close an analogy with the rings of Saturn. Secondly, we must consider the time-scale. There could be no chance of the Moon coming within the Roche limit for well over 10,000 million years (probably 30,000 million years is more realistic), and by that time the Earth-Moon system as we know it will no longer exist, because of changes in the Sun.

This is no place to discuss stellar evolution in any detail, but we may be sure that the Sun will not last for ever. It is not 'burning' in the conventional sense; it is shining because it is changing one element (hydrogen) into another (helium), losing mass in the process. The mass-loss amounts to four million tons per second, which seems staggering, but is negligible when we remember how much material the Sun contains. When the supply of hydrogen 'fuel' runs low, the Sun will have to

39

readjust itself. The core will shrink and heat up; the outer layers will expand and cool down, and the Sun will become a Red Giant star, as Betelgeux in Orion and Antares in the Scorpion are now. There will be a period when it will increase its output of energy by a factor of at least one hundred, and this will be fatal to the inner planets. The crisis will not be upon us for at least 5,000 million years, and probably longer, but it will certainly come before the Moon returns to anywhere near the Roche limit.

All in all, it seems that the Moon will survive for as long as the Earth itself. We are dealing with events so far in the future that speculation is really rather pointless, but we may be confident that nothing dramatic will happen to the Moon yet awhile.

There was never any serious suggestion that the lunar voyages would reveal any unknown substances on the Moon, and the rocks brought back for analysis in our laboratories are made up of familiar materials—though in some cases rather unfamiliar in structure, because of the different conditions under which they have evolved. In fact, the Earth and the Moon are of essentially the same nature. The superficial differences are due mainly to the Moon's lower density and lesser mass.

If I am right in suggesting that the Moon is a companion planet rather than a true satellite, what can we say about the attendants of other planets? I suspect that their origins may be rather different even for the large members of the families of the giants, and there is undoubtedly a fundamental difference between the Moon and the smaller satellites of Jupiter, Saturn and Neptune, to say nothing of the dwarf moonlets of Mars. All these may be captured asteroids. We have reliable information about both Phobos and Deimos, because they were photographed from close range by Mariner 9; there is nothing lunar about them, except that they too are crater-scarred. From Earth they appear as faint specks of light, but they are not so hard to see as might be thought. Under good conditions, I have been able to glimpse them with the modest 15-inch reflecting telescope in my own observatory.

I make this point because it brings me on to the interesting question of whether the Earth can have any small asteroid-type satellites. Some years ago there was considerable popular interest in Toro, one of the asteroids, which was said to be really a second Earth satellite, but in fact this is wrong. Toro moves round the Sun in the usual way, though it does make periodical close approaches to us ('close', that is to say, in the cosmical sense; it never comes within two million miles of us, and it is very small, so that it is much too faint to be seen with the naked eye).

There is certainly no junior satellite within reasonable range. At a distance equal to that of the Moon, a 25-mile body of average reflecting power would shine as a brilliant star, and a satellite only one mile across would be visible in binoculars even if it were made up of darkish rock. At 2,000,000 miles, a 25-mile body would be seen with the naked eye, and a one-mile satellite would be detectable with equipment of the type used by the average amateur. Photographic surveys of the sky are now so complete that we may be sure that anything of this sort would have been tracked down years ago. If a minor satellite exists, it must be either much further away or else very diminutive indeed.

In the nineteenth century, a French writer named Petit published a paper about a suggested second satellite moving at a distance of 4,650 miles from the Earth's surface, and with a period of 3 hours 20 minutes. I am not sure how Petit arrived at this conclusion, which is certainly wrong, but the idea was taken up by Jules Verne in his famous novel *Round the Moon*, published more than a hundred years ago. I will say more about Verne later; meanwhile it is worth repeating his description of what was seen by one of the fictional space-travellers, President Barbicane, observing through the porthole of the equally fictional moon-projectile:

As Barbicane was about to leave the window . . . his attention was attracted by the approach of a brilliant object. It was an enormous disk, whose dimensions could not be estimated. Its face, which was turned earthward, was brightly illuminated; it might have been taken for a small moon reflecting the light of the large one. It advanced very rapidly, and seemed to be following an orbit round the Earth which would interest the path of the projectile. It moved along in its orbit, and at the same time spun on its own axis . . . The object grew enormously, and the projectile seemed to be rushing into its path . . . The travellers instinctively recoiled. Their alarm was great, but it did not last long. The object passed within a few hundred yards and vanished, merging into the absolute blackness of space.

It is a wonderful description, and the second satellite was essential in the ingenious plot of Verne's story, but it was nothing more than that. Quite recently Clyde Tombaugh, the American astronomer who made history in 1930 when he tracked down the ninth planet, Pluto, made a really systematic search for minor Earth satellites, and found none. Tombaugh, who made use of the powerful telescopes at the Lowell Observatory in Arizona, has told me that he is sure there can be no natural orbiting body worthy to be ranked as a satellite. Of course, there may well be tiny bodies moving round us at distances less than that of the Moon, but they must be so small that they must be classed as meteoritic débris.

There is, incidentally, an intriguing suggestion made by the Polish astronomer K. Kordylewski. It was first announced in 1961, and is still unconfirmed, but it is worth noting.

Kordylewski began his search for minor satellites in 1951, using the equipment at Kasprowy Wierch and Lomnica in the mountains of Poland. On 6 March and 6 April, 1961 he took photographs which, he claimed, showed two faint 'clouds' moving in the same orbit as the Moon, and presumably made up of meteoritic material. The search had not been a random one; Kordylewski had been studying these special areas, because of the connection with the famous 'three-body' problem. To explain this, we must look at the remarkable asteroids known as the Trojans.

The Trojans lie far beyond the main asteroid belt, and move in the same orbit as Jupiter, at a mean distance of about 483,000,000 miles from the Sun. There are two groups of Trojans, one of them 60 degrees ahead of Jupiter and the other 60 degrees behind. They do not, of course, keep together in clumps, and may be spread out by many millions of miles, but the 60-degree points represent their mean positions, and there is no fear that any of them will wander dangerously close to Jupiter. They keep well out of harm's way. In the diagram, A represents Jupiter, and B and C the two Trojan groups, with S marking the position of the Sun.

42

Kordylewski suggested that the same could be true of his 'clouds'. In Fig. 6, S now stands for the Earth and A for the Moon, while the 'clouds' are at B and C. The cloud at B would keep 60 degrees behind the Moon, and the cloud at C 60 degrees ahead.

Up to now nobody has been able to confirm the existence of the clouds, but even if they exist they are certainly not satellites, and can be nothing more than loose collections of tiny particles. No doubt we will eventually decide whether they are real or not.

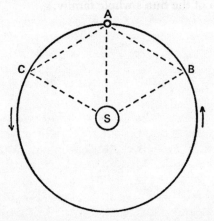

Fig. 6. Position of the Trojans relative to Jupiter, and also Kordylewski's reported objects relative to the Moon

In passing, it is worth referring back to a much older suggestion—dating back for centuries—that there may be a second Earth satellite lying directly behind the Moon, so that it is permanently hidden from us. This is an attractive idea, but it is completely unsound. The perturbations produced by other bodies in the Solar System would soon destroy the exact lining-up, and the satellite would come into view. (For the same reason, there is no possibility of an extra planet moving in the same path as the Earth and keeping behind the Sun.) We must assume that the Moon is the Earth's only natural companion, though by now there are plenty of artificial ones.

Finally, what are the chances of a satellite of the Moon

itself? Seventy years ago W. H. Pickering made a careful search for one. He failed completely, and later photographic surveys by Clyde Tombaugh have been no more fruitful. Tombaugh has concluded that there can be no natural lunar satellite with a diameter of more than fifteen feet or so.

I fear that this has been something of a digression, but the whole question is of tremendous interest, and at least we can hope that future studies from the Moon itself will help in clearing up some of the puzzles about the history of the Earth-Moon system. This, in turn, will tell us more about the origin and evolution of the Sun's whole family.

Chapter Five

THE MOVEMENTS OF THE MOON

'THE MOON MOVES ROUND the Earth.' This is the bald statement given in countless books, and it is good enough for most purposes, even though the scientist will need to qualify it. There has never been any doubt that the Moon is our companion, and that it is the nearest natural body in the sky. The Greeks knew this; so, presumably, did many of their predecessors. Where they went wrong was in supposing that the Sun, planets and stars must also circle the Earth.

Obviously the most noticeable thing about the Moon is that it seems to change in shape. Sometimes it shows up as a crescent, sometimes as a half-disk and sometimes as full, while there are periods when it cannot be seen at all. Even today there are some people who have strange ideas about these 'phases', but there is nothing in the least mysterious about them. To make things as straightforward as possible, let us imagine that the Moon is travelling round the Earth once a month in a perfectly circular orbit. This is shown in the diagram, which, like those following, is wildly out of scale.

The Moon has no light of its own, as the Greeks knew quite well. It shines only by reflected sunlight, and clearly the Sun can illuminate only half of the Moon at any one time. In the diagram (Fig. 7), the unlighted and therefore non-luminous hemisphere is blackened, while the shining or day-hemisphere is left white; E represents the Earth, S the Sun, and M1, M2, M3 and M4 the Moon at various positions in its orbit.

Look first at M1. At this moment the Earth, Moon and Sun are in more or less a straight line, with the Moon in the mid position. The lighted half is (as always) turned toward the Sun, which means that the dark half is facing the Earth, and the Moon is new. People often speak of the thin crescent moon as being 'new', but this is scientifically wrong; the true new moon cannot be seen at all.

From M1, the Moon moves along in the direction of M2. Gradually a little of the day-hemisphere begins to show, and

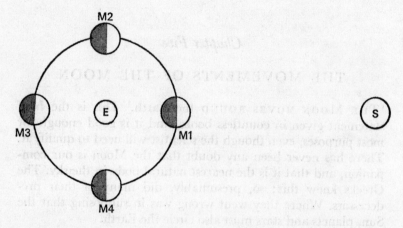

Fig. 7. Phases of the Moon

the familiar crescent makes its appearance in the evening sky; this is the best time to see the earthshine, which makes the night-hemisphere dimly luminous. By the time M2 is reached, half of the day-hemisphere is visible from Earth, and the Moon is said to be at First Quarter.

It may sound odd to speak of First Quarter when the Moon appears as a half-disk, but there are good reasons for it. The Moon has completed one-quarter of its journey, reckoning from new to new; also, we are seeing one-quarter of the total surface.

From M2 the Moon moves on toward M3, and more and more of the day-side comes into view. The Moon is 'gibbous', i.e. between half and full. At M3 the whole of the sunlit hemisphere faces us, and the Moon appears as a complete disk. Once again Earth, Moon and Sun are more or less lined up, but this time the Earth is in the mid position.

As the Moon continues on its way toward M4, the day-side begins to turn away from us again. Passing through the gibbous stage, the Moon has become a half-disk by the time it arrives at M4—the phase known as Last Quarter—and it again approaches the Sun's line of sight, becoming a narrowing crescent and finally disappearing into the morning dawn. After $29\frac{1}{2}$ days it has come back to M1, and is again new.

There is an obvious discrepancy here. The Moon takes only

27·3 days to go once round the Earth; why then is the interval between successive new moons over two days longer?

The reason is that the Earth itself is moving round the Sun—see Fig. 8, where we have the Sun at S, the Moon at M, and the Earth in two different positions, E1 and E2. When the Earth is at E1 and the Moon at M1, the Moon is new. After 27·3 days it has completed one circuit, and has arrived back at M1; but meanwhile the Earth has moved on to E2, and the Moon must travel further along its orbit, to M2, before the three bodies are properly lined up again. The extra time taken to cover the distance between M1 and M2 is just over two days, which accounts for the discrepancy.

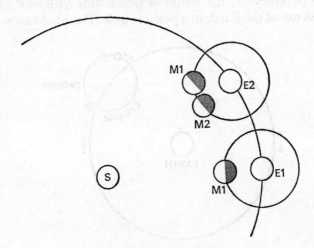

Fig. 8. The Lunation

There are technical terms for these periods. The 27·3-day revolution time is known as the Moon's sidereal period, while the interval between one new moon and the next is the 'lunation' or synodic month.

The next correction has to do with the shape of the Moon's path, which is definitely not circular. The Greeks were basically right in saying that the Moon moves round the Earth, but in almost every other respect they were hopelessly wrong. The root cause of the trouble was their utter faith in circular orbits.

47

Astronomers of ancient times, including Ptolemy, held that the circle is the 'perfect' form, and this meant that all the celestial bodies would have to have circular paths, since nothing short of absolute perfection could be allowed in the heavens. Unfortunately for this theory, it was obvious from the outset that the Moon does not move in a completely regular way, and neither does it always look the same size, so that its distance from the Earth must vary. One way round the difficulty would be to assume that the Earth lies some distance from the centre of the Moon's circular orbit, but even this would not account for what is actually observed. Ptolemy, who was an excellent mathematician, decided that the Moon must move in a small circle or 'epicycle', the centre of which—the 'deferent'—itself moved round the Earth in a perfect circle (Fig. 9). As more and

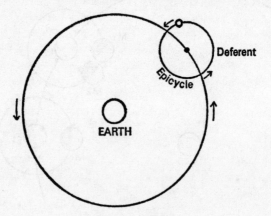

Fig. 9. Epicycles

more problems arose the number of epicycles had to be increased, so that the whole scheme became hopelessly involved and artificial.

After Ptolemy's time there was a long period of stagnation, and what we may call the modern era of astronomy dates only from the publication of Copernicus' great book in 1543. But though Copernicus was on the right track, he too was wedded to the notion of perfectly circular orbits, and he was even reduced to bringing back epicycles. The final solution was left to Johannes Kepler, whose first Law, published in 1609, stated that the planets move round the Sun in ellipses. The same is true of the Moon; it moves in an elliptical orbit, with the Earth in one of the foci.*

This explains the variation in the Moon's apparent size (Fig. 10). At its closest to us, or 'perigee', the distance from the centre of the Earth is 221,500 miles; at its furthest, or 'apogee', the Moon recedes to 252,700 miles, giving a mean of rather less than a quarter of a million miles. The changes in distance are quite considerable, and the Moon's apparent diameter at apogee is only 9/10 of the value at perigee. The difference is not marked enough to be noticeable with the naked eye, but it is easy to measure. On the other hand it would be most misleading to think of the lunar orbit as being highly eccentric. If it were drawn to a scale of, say, three inches in diameter, so that it could be fitted on to a page of this book, it would look circular unless carefully measured.

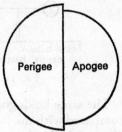

Fig. 10. Apparent size of the Moon at perigee and at apogee

The next correction is of different type, and brings me back to my contention that the Earth-Moon system should be classed as a double planet rather than as a planet and a satellite. Therefore, we cannot properly say that the Moon moves round the Earth. What happens is that both bodies move round their common centre of gravity.

Consider an ordinary gymnasium dumbbell (Fig. 11). Balance it on a post by its joining arm, and twist it; both bells will revolve round the centre of gravity of the system, i.e. the point where the arm is supported. Ordinarily this point will be in the middle of the arm, since the bells will be equal in weight. If one bell is heavier than the other, the balancing point will be closer to the heavier bell; and the greater the difference in

* The accepted way of drawing an ellipse is to fix two pins in a board, an inch or two apart, and fasten them to the ends of a length of cotton, leaving a certain amount of slack. Then draw the cotton tight, and trace a curve, keeping the cotton tight all the time. The result will be an ellipse, with the pins marking the foci. In practice, of course, what always happens is that the thread breaks and the pins fall out.

weight, the greater will be the distance of the supporting point from the centre of the arm.

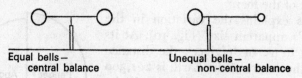

Equal bells—   Unequal bells—
central balance   non-central balance

Fig. 11. Centre of gravity demonstration

The same holds good for the Earth and Moon, which may be compared with the two bells, with the force of gravity taking the place of the joining arm. The Earth has 81 times the mass of the Moon, and so the centre of gravity is shifted earthward—so far, in fact, that lies inside the terrestrial globe, though some way from the middle of the Earth. It is around this balancing-point, or barycentre, that the two bodies are revolving.

Even now there is yet another correction to be made to our original diagram, which showed the Moon moving round the Earth in a conveniently circular orbit. This new complication arises because, strange though it may seem, the Sun's pull upon the Moon is more than twice as powerful as the Earth's.

The Earth is moving round the Sun at a mean velocity of $18\frac{1}{2}$ miles per second, or roughly 66,000 m.p.h. Relative to the Earth, the Moon's orbital velocity is only about one-third of a mile per second, so that it too is moving at approximately 66,000 m.p.h. relative to the Sun; and to an observer in space the Moon must appear as a normal planet, travelling in an elliptical orbit with the Sun in one of the foci. The path of the Moon is always concave to the Sun, as shown in Fig. 12.

Although the Sun's pull is so strong, there is no chance that the Moon will part company with the Earth and move away on its own. This is because the Sun attracts the Earth and Moon almost equally. The Moon is the more strongly pulled when it lies between the Sun and the Earth, around the time of new moon, and less strongly when it is on the far side, and is full; but the force on the two bodies is always much the same.

Next, let us turn to eclipses, which have been carefully studied for many centuries, and which caused a great deal of alarm and despondency before astronomers found out why they happen.

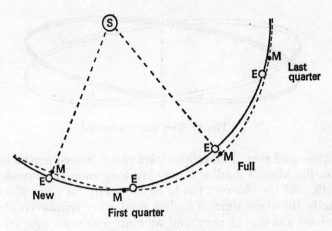

Fig. 12. The Moon's orbit always concave to the Sun

Although the Sun is so much larger than the Moon, it is also so much further away that it appears almost exactly the same size in the sky; in each case the angular diameter is about half a degree. Consequently when the Earth, Moon and Sun move into a line, with the Moon in the middle, the lunar disk blots out the Sun, and we have a solar eclipse. If the Moon's orbit were really as simple as shown in the first diagram, there would be a solar eclipse at every new moon. This does not happen, because the Moon's path is tilted at an angle of approximately five degrees to the plane of the orbit of the Earth.

The apparent yearly path of the Sun among the stars marks what we term the ecliptic, and can be calculated very accurately, even though the stars cannot be seen with the naked eye when the Sun is above the horizon (except during the fleeting moments of a total solar eclipse). The monthly path of the Moon across the sky is easy to chart, and proves to be inclined relative to the ecliptic. One way to show this is to compare the orbits with two hoops (Fig. 13), hinged along a diameter and placed at an angle to each other, as shown here. The tilted hoop will lie half above and half below its companion, and this is also the case with the Moon's apparent path compared with that of the Sun. The two points where the 'hoops' cross are termed the nodes.

Unless new moon falls near a node, there can be no solar

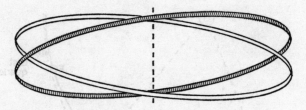

Fig. 13. Two inclined hoops

eclipse—and even if an eclipse takes place, it need not be total. Also, the Moon's shadow is only just long enough to touch the Earth, and the observer has to be in exactly the right place at exactly the right time. The last total solar eclipse visible in England was that of 1927, and we must wait until 1999 for the next, though other parts of the world are more favoured. Incidentally, no total solar eclipse can last for as long as eight minutes, and most are a great deal shorter. I once went to Siberia to see a total eclipse which lasted for a mere 37 seconds.

Lunar eclipses are quite different. They take place when the Moon passes into the shadow cast by the Earth, and the supply of direct sunlight is unceremoniously cut off. I will describe them more fully in Chapter 11. They, too, depend upon the lining-up of the Moon, Earth and Sun, and can take place only when the Moon is full. More than a dozen lunar eclipses will be visible from England before the end of the century.

The Moon's orbit is not absolutely unchanging for revolution after revolution. The effects due to the gravitational pulls of the Sun and Earth change, and there are other factors to be taken into account as well. The result is that the nodes shift slowly round the orbit, completing a full circuit in just over eighteen years.

Eclipses merit a chapter to themselves, so for the moment let me turn to another famous phenomenon—Harvest Moon.

Everyone knows that the Moon rises in an easterly direction and sets toward the west. This is due to the real rotation of the Earth on its axis, from west to east. The Moon is also moving in its orbit from west to east, and so it seems to travel eastward among the stars, covering about thirteen degrees per day. Anyone who takes the trouble to check the position of the Moon

52

on successive nights, using nearby stars as reference points, will soon see just what is happening.

The apparent path of the Moon in the sky is not very different from that of the Sun; the angle between the two is only five degrees, which is not very much even though it is sufficient to prevent eclipses from happening every month. When full, the Moon is opposite to the Sun in the sky, and to observers in the northern hemisphere of the Earth it lies due south at midnight.

In Fig. 14, the angle of the ecliptic compared with the horizon is shown for spring and for autumn. (Again I am speaking from the viewpoint of northerners; if you happen to live in South Africa or Australia, everything is reversed.) In spring, around March, the angle is at its steepest. In 24 hours

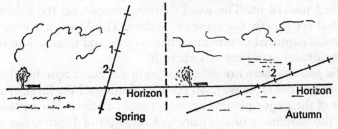

Fig. 14. Harvest Moon

the Moon moves from position 1 to position 2, and obviously the 'retardation'—that is to say, the difference in rising-time from one night to the next—is considerable. The situation in autumn is different. The angle is much shallower, and although the Moon moves against the starry background by the same amount—in other words, the distance between 1 and 2 is the same in each diagram—the retardation will be less. In September, the retardation may be reduced to about a quarter of an hour, no matter where you happen to be on the surface of the Earth.

It is often said that in September the full moon rises on the same time on several successive evenings. This is not true, and the retardation is always appreciable, but it is noticeably less than at other times of the year. The September full moon is called Harvest Moon, because farmers used to find it very useful as a source of extra light at a particular busy time. The following full moon (Hunter's Moon) behaves in much the same way, though the retardation is greater.

It is also been claimed that Harvest Moon and Hunter's Moon look particularly big. This is quite wrong—they appear the same size as other full moons; but this leads on to the whole fascinating problems of the so-called Moon Illusion, which I propose to discuss in more detail in Chapter 6. Meanwhile, let us go on to the question of the Moon's axial rotation.

I have already touched upon the subject, when talking about the tides raised by the Earth in the viscous body of the young Moon. As we have noted, the Moon's rotation was strongly braked, until relative to the Earth (though not relative to the Sun or the stars) it had stopped altogether. For thousands of millions of years now, the Moon has kept the same hemisphere turned toward us. The axial rotation period and the sidereal period are exactly the same: 27·3 days. The Moon has what is called a captured or synchronous rotation, and there is nothing coincidental or mysterious about it.

En passant, there are other bodies in the Solar System which behave in the same way. So far as we know, all the large satellites of the major planets have captured rotations. For instance Io, the innermost of the principal satellites of Jupiter, has an orbital period of 1·8 Earth-days, and its axial rotation period is the same; with Iapetus, the outermost of the senior attendants of Saturn, the period is as much as 79 Earth-days, and so on. The same is true for Phobos and Deimos, the midget satellites of Mars. It used to be thought that the planet Mercury, with its orbital period of 88 days, behaved similarly—in which case one hemisphere would have been permanently sunlit and the other plunged into eternal night. Science fiction writers made great play of this odd state of affairs, but recent investigations have shown that Mercury's rotation is not captured; it amounts to only 58½ days, so that all parts of the planet are in sunshine at some time or other.

It has often been claimed that the Moon does not spin at all, and over the years I have had many letters about it. I have never understood why, because a very simple experiment (Fig. 15) can show that the idea is absurd; we would see all parts of a non-rotating Moon. Put a chair in the middle of the room to represent the Earth, and imagine that your head represents the Moon. Stand behind the chair, a foot or two away from it, and fix your eyes upon some object beyond, such as a picture on the wall. Now walk in a circle round the chair,

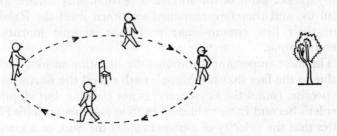

Fig. 15. Chair demonstration

keeping your eyes fixed on the picture all the time. When you have completed half your circuit, you will find that the picture is in front of you and the chair behind, so that your back hair, not your face, is pointing chairward. To keep your face turned toward the chair all the time, you must turn as you walk; in other words you must rotate upon your axis, completing one revolution for each trip round the chair.

In fact, the Moon keeps the same hemisphere turned toward the Earth, but not toward the Sun. From Earth we can never see the 'other side' of the Moon, but conditions of day and night there are the same as on the familiar hemisphere.

Look at the Moon, even with the naked eye, and you will see obvious features; the dark plains are striking, and any pair of binoculars will show vast numbers of craters. The positions of these features on the disk are always much the same. For instance, the well-marked plain which we call the Mare Crisium is always to the upper right as seen with the naked eye (as seen from the northern hemisphere), while the dark-floored crater Grimaldi is to the mid-left, and the 90-mile walled plain Ptolemæus almost in the centre. From Earth you will never see these features, or any others on the Moon, in different positions. Once you have recognized the various formations, you will find no difficulty in identifying them again, because their positions are unaltered—or nearly so.

I say 'nearly so' because there are effects known as librations which do cause slight modifications, and from Earth it is possible to examine a grand total of 59 per cent. of the Moon's surface instead of exactly 50 per cent., though needless to say we can never see more than 50 per cent. at any one moment. Only 41 per cent. of the surface is permanently turned away from us, and therefore remained unknown until the Russians sent their first circum-lunar probe on its epic journey in October 1959.

The most important libration—the libration in longitude— is due to the fact that the Moon's path round the Earth (or, to be precise, round the barycentre) is not circular, but elliptical. Kepler's Second Law, published in 1609 together with his First, states that the velocity of a planet round the Sun, or a satellite round its primary, depends upon its distance. The rule may be summed up neatly by the phrase 'the nearer, the faster'. Thus Mercury, at a mean distance of 36 million miles from the Sun, moves much more quickly in its orbit than the Earth, at 93 million miles.

The Law is equally valid for the Moon, where the distance from Earth ranges between about 221,000 miles at perigee to about 253,000 miles at apogee. When near perigee it is moving at its fastest, while by the time it reaches apogee it has slowed down. In fact, the Moon's velocity in orbit is not constant, whereas the rate of axial rotation does not change at all—a situation which has far-reaching results.

The next diagram (Fig. 16) should make things clear. Begin with the Moon at perigee, in position 1, and take X as marking the centre of the disk as seen from Earth. After a quarter of its journey, the Moon has reached position 2; but since it has travelled from perigee it has moved slightly quicker than its mean rate, and has covered 96 degrees instead of only 90. As seen from Earth, position X lies slightly east of the apparent centre of the disk, and a small portion of the 'far side' has come into view in the west, so that in fact we are seeing a short way beyond the mean western limb.

After a further quarter-month the Moon has reached position 3. It is now at apogee, and point X is again central. A further 84 degrees is covered between positions 3 and 4, and X is displaced toward the west, so that an area beyond the mean eastern limb is uncovered. At the end of one revolution the

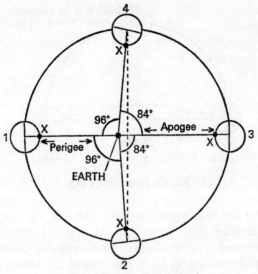

Fig. 16. Libration in longitude

Moon has arrived back at 1, and X is once more central on the disk as seen from Earth.

Libration in longitude means, then, that the Moon seems to wobble slightly to and fro, allowing us to peer alternately beyond the mean east and west limbs of the Earth-turned hemisphere. There is also a libration in latitude; because the Moon's orbit is perceptibly tilted, we can sometimes see for some distance beyond the mean northern or southern limb. Finally there is a diurnal or daily libration, due to the axial rotation of the Earth itself; at moonrise an observer at A (Fig. 17) can see slightly beyond the mean western limb, and at moonset slightly beyond the mean eastern limb.

The sum total of all these effects is that at one time or another we can study 59 per cent. of the surface, but the so-called libration zones, i.e. those which are periodically brought into and out of view, are difficult to map from Earth, because all the features are seen at inconvenient angles and are badly foreshortened. Before the Space Age our charts of these regions were decidedly uncertain—as I well know, since I spent over twenty-five years in trying to map them. Luna 3, the Russian probe of 1959, marked the beginning of a new era, since it went

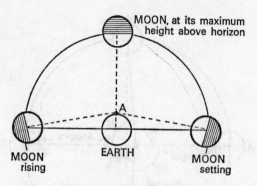

Fig. 17. Diurnal libration

right round the Moon and sent back the first pictures of the far side, confirming that there too the scene is dominated by mountains and craters. Since then various other probes, manned and unmanned, have gone round the Moon, so that by now we have very accurate charts of virtually the entire surface.

This seems the right moment to refer briefly to another peculiarity of the lunar motion: the secular acceleration, or apparent speeding-up of the Moon in its orbit. If we take the position of the Moon as measured centuries ago, and then predict the present position by adding on the correct number of revolutions, the results will not agree; the Moon will have moved too far—or, to all appearances, too quickly.

Fortunately this can be checked, because ancient astronomers have left us useful records of eclipses. A total eclipse of the Sun can happen only at new moon, and therefore the moment of totality is also the exact time of new moon. Eclipse records date back for thousands of years, and it was by comparing these records with modern measurements of the Moon's position in the sky that the apparent speeding-up was discovered. It is due partly to Venus and Mars, which pull on the Earth and make our orbit round the Sun slightly less elliptical, and partly by the tidal effects described in the last chapter. The effect is very slight, but over many hundreds of years it mounts up sufficiently to be appreciable. I will return to the problem in Chapter 11.

I appreciate that this description of the Moon's movements is very over-simplified; the complications are severe enough to tax the ability of the world's greatest mathematicians. We have learned a great deal since the men of ancient civilizations looked up at the Moon and wondered why it showed its monthly changes in form.

Chapter Six

THE MOON AND THE EARTH

UNIMPORTANT THOUGH THE MOON may be in the Solar System as a whole, it affects us on Earth more than any other celestial body apart from the Sun. Without the Moon, life would seem strange indeed. We could manage without its light—as we have to do for part of every month—but what about the tides?

I happen to live on the coast, within sound of the sea, so that I always know whether the tide is in or out; but there can be nobody who is not familiar with the ebb and flow of the waters, and it is obvious that the tides are associated with the Moon. The Sun is concerned too; but for the moment let us consider only the Moon—because the theory of tides is much more complicated than might be thought, and it is best to separate the various factors as much as possible.

As a start, imagine that the whole Earth is covered with a shallow, uniform ocean, and that both the Earth and the Moon are standing still. In Fig. 18 (as usual, hopelessly out of scale) we have a high tide at point A, and another high tide at point B, with low tides at C and D. Basically, the Moon's gravitational pull is heaping the water up at A, where the force is strongest. Of course the pull of gravity affects the solid land also, but land is more difficult to move about than water, so that the heaping-up is less.

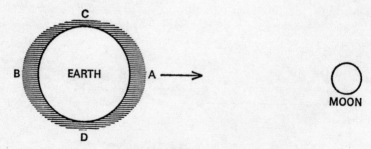

Fig. 18. Theory of the Tides

This is all very well, but at first sight it is not so easy to see why there should be a second high tide at B, on the far side of the Earth. It is slightly misleading to say, as many books do, that the solid globe is being simply 'pulled away' from the water, so for a moment let us assume that we have a situation in which the Earth and the Moon are absolutely alone in the universe, and are falling toward each other because of their mutual gravitational pulls. Taking matters a step further, imagine that it is the Moon which is standing still, and the Earth which is being drawn toward it. Point A, which is closest to the Moon, will be subject to a greater accelerating force than the average, and so the water will bunch up around it, though the pull of the Earth will prevent the water from leaving the solid surface. The result will be a high tide. At B, the reverse will happen. The acceleration will be less than the average, and so the water in that region will tend to get 'left behind', so bulging away and causing a similar high tide.

Next, assume that the Earth and the Moon are keeping the same distance from each other, but that the Earth is spinning round once in 24 hours. Obviously, the water-heaps—that is to say, the high tides—will not spin with it; they will keep 'under' the Moon. Therefore, each bulge will seem to sweep right round the Earth once in 24 hours, and every region will have two high tides and two low tides per day.

Now we can start to introduce a few of the many complications. First, the Moon is not stationary; it is moving along in its orbit, so that the water-heaps shift slowly as they follow the Moon around. On average, the high tide at any particular place will be 50 minutes later each day. Neither will the two daily high tides be equal. Consider a point C (Fig. 19), which has a high tide. Twelve hours later it will have moved round to C¹, and will have another high tide—but this high tide will not

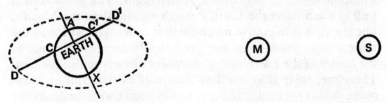

Fig. 19. Diagram to show the 'diurnal inequality' in the tides

be so great as the first, because the Earth's axis, AX, is tilted with respect to the path of the Moon. If the original tide is represented by CD, the second will be given by C^1D^1, which is decidedly less. This difference is known as the 'diurnal inequality' of the tides.

But, of course, the Earth is not surrounded by a shallow, uniform water-shell. The seas are of various shapes and depths, and local effects are all-important. At Southampton, for instance, two high tides occur in succession as the rising water comes up the two narrow straits separating the mainland from the Isle of Wight, first up the Solent and then up Spithead. In the Bay of Fundy in Nova Scotia, and also along some coastal areas in South America, the tidal range is fifty feet or so, while in other places it is less than twelve inches. Actually, the forces involved are surprisingly small, and in mid-ocean the tides do not exceed a couple of feet. The majestic tides of the Bay of Fundy class are due to this deep-ocean 'wave' washing up on the shallow shelf of the coast, or being concentrated by converging coastlines.

Also, the waters take some time to heap up, and maximum tide does not occur directly under the Moon. There is an appreciable lag, and the highest tide follows the Moon after an interval which varies according to local conditions. Naturally, the lag is usually greatest for shallow coastal areas.

The next complication involves the changing distance of the Moon. Near perigee, when the Moon is at its closest to the Earth, the pull is stronger than at the time of apogee, and the tides are higher; in fact the difference amounts to over 20 per cent., though again it is impossible to give a consistent figure because of the different conditions over various regions of the Earth.

Next we come to the Sun—and here we are faced with a situation which seems curious at first sight. The gravitational pull of the Sun on the Earth is much greater than the Moon's, but the Sun is of course much further away, and so far as the tides are concerned what matters is the difference of the pull at the centre of the Earth and at the point directly under the Sun. Therefore, solar tides are less than half as powerful as lunar ones; in this particular tug-of-war the Moon wins easily. What we have to do is to consider the two tide-raising bodies together.

At new or full moon—the positions when the Moon is said to be at 'syzygy'—the Sun and the Moon are pulling in the same direction (or in exactly opposite directions, which comes to the same thing) and produce the strong tides known, rather misleadingly, as spring tides. The much lower neap tides occur when the Moon is at 'quadrature', i.e. First or Last Quarter, so that its gravitational pull works against that of the Sun. This may sound rather involved, but I hope that the diagram (Fig. 20) will make the situation clear.

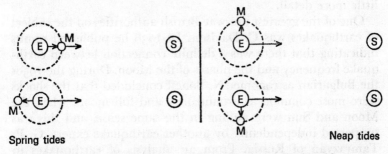

Spring tides   Neap tides

Fig. 20. Relative positions of Sun and Moon affecting tides

The Moon pulls upon the solid body of the Earth as well as upon the oceans, and land tides are easy to measure with suitable instruments. A rather odd fact has emerged from studies of them. Although the body of the Earth behaves as though it were more rigid than steel, it also proves to be perfectly elastic; after being distorted by the tidal pull it returns to its original shape without any appreciable delay, rather in the manner of an elastic band which is first stretched and then let go. However, land tides amount to only about $4\frac{1}{2}$ inches instead of many feet, and in everyday life they are too minor to be noticed at all.

There are tides in the atmosphere, too, and there may possibly be some effect upon radio reception. Long-distance wireless communication is made practicable by the presence of reflecting layers in the upper part of the atmospheric blanket, known as the ionosphere. As long ago as 1939 two British scientists, Appleton and Weekes, showed that the Moon causes tides in the ionosphere, and some researchers believe that radio

63

reception is best when the Moon is near syzygy. Others have serious doubts. Observations of this kind are very difficult, and it is far from certain that the Moon has any real influence upon radio conditions. In any case, it is very slight.

There is no doubt that some parts of the Earth's interior are liquid, so that tides must be produced there also, and it has even been suggested that the tides inside the Earth may have some connection with earthquake shocks. Again we are on very uncertain ground, but since I have carried out a few mild investigations on my own account I feel justified in going into a little more detail.

One of the greatest pre-war British authorities on the subject of earthquakes was C. Davison. In 1938 he published results indicating that there was a definite connection between earthquake frequency and the phases of the Moon. During the 1950s the Bulgarian astronomer N. Boneff concluded that the shocks were most common at around new and full moon, when the Moon and Sun were pulling in the same sense, and this was supported independently by another earthquake expert, G. P. Tamrazyan of Russia. From an analysis of earthquakes in California up to the end of 1935, Maxwell Allen had already proposed some sort of lunar control.

It all seemed rather nebulous, so I decided to take a close look at the world-wide records which are collected at Kew Observatory in London. Unfortunately I could find no reliable lists showing the violence of individual earthquake shocks, but some sort of guide is given by the number of recording stations recording each earth tremor; an earthquake recorded by, say, a hundred stations is likely to be much stronger than a shock reported by only half a dozen. On this basis, I took the Kew records for the period from 1931 to 1953 and worked out an 'earthquake activity number' for each day, after which I tried to correlate these numbers with the phases of the Moon. The results were absolutely negative; so far as I could make out, there was no connection at all. Trying to link the earthquakes with lunar perigee and apogee proved equally fruitless. Of course, the whole investigation was very rough, but so far as it went it did not support the idea of any lunar control. It is best to say that the whole question remains open. It is true that many earthquakes have occurred at full moon—but the Moon is full once a month, and a search for coincidences will nearly always reveal them.*

As a minor digression, it may be worth commenting upon another idea which has been widely publicized of late. During the early 1980s the outer planets will be roughly lined up, so that they will be pulling in the same direction, and there have been sensational predictions that this will cause a major earthquake in the San Francisco area of California. Actually, the forces involved are so tiny that I for one have not the slightest faith in anything of the kind. The planets are more massive than the Moon, but are also so much further away that as earthquake-raisers we can undoubtedly disregard them. There may well be another San Francisco earthquake before long, and seismologists regard it as only too likely, but to blame the planets for any such disaster is, in my view, far-fetched.

The Moon and the weather are often said to be linked, but there is no conclusive evidence for any connection between them, and attempts to associate the lunar phases with rainfall, droughts, changes of temperature and the like are, to put it mildly, unconvincing. Admittedly the weather does often change at full moon, but, after all, the weather changes every

* I could give dozens of examples. For instance, in 1952 an American radio engineer published a paper in which he claimed that certain planetary configurations have a marked effect upon radio reception. The frequency-curves in his paper looked most impressive at first sight, but it was clear that he had been forced to bring in so many 'configurations' that he was, no doubt unintentionally, coincidence-hunting. When his paper was described at a meeting of the British Astronomical Association, it was shown that his configurations were even more closely related to (1) the light-fluctuations of the variable star Delta Cephei, and (2) the frequency of matinée performances at the Folies Bergère in Paris. With regard to (1), Delta Cephei is 600 light-years from the Earth, so that its rays now reaching us started on their journey about the time of the Crusades, and it does not seem very likely that they can have much to do with the present positions of the planets in their orbits—yet the frequency-curves fitted in very well. I am hardly qualified to comment upon case (2), since I have not been to the Folies Bergère since I last attended a scientific conference in Paris some years ago.

Coincidence-hunting can be quite amusing, though completely useless. I have no doubt that by selecting suitable qualifications it would be possible to draw graphs showing links between—say the periods of the minor planets, the price of bananas, and the number of goals scored by the Aston Villa football club. Astrologers are particularly good at this sort of thing.

few days—in England, at least—so that we are back to co-incidence-hunting. An old country saying tells us that 'the full moon eats up the clouds', and certainly the sky sometimes clears when the full moon rises; but at the same time, the Sun must be setting—and nobody can deny that weather and cloud conditions are very strongly dependent upon the Sun! Thunderstorms and meteors have also been linked with the Moon's phases, but again without justification.

All the same, it is worth saying a little about some of the atmospheric effects seen during moonlight, even though they belong strictly to the realm of meteorology and have no actual connection with the Moon in the astronomical sense.

Everyone knows the expression 'once in a blue moon', but how many people have really seen a blue moon? I have—once, on 26 September, 1950, when I was living at East Grinstead in Sussex. Late in the evening I recorded that 'the Moon shone down from a slightly misty sky with a lovely shimmering blueness—like an electric glimmer, utterly different from anything I have seen before'. The phenomenon was certainly not confined to East Grinstead. Over a period of at least forty-eight hours, people in various parts of the world were startled to see not only a blue moon but also, in some places, a blue sun, and some of the Press reports were highly sensational. Yet there was no mystery about it. Giant forest fires raging in Canada had sent a tremendous amount of fine dust into the upper atmosphere, and it was this dust which caused the eerie colouring. From all accounts, the dust-pall over North America was really remarkable. At Ottawa, car headlights had to be switched on at midday, even though there was no fog, and in New York a game of baseball was played under arc-lights. Other blue moons have been seen occasionally, always for the same sort of reason.

Haloes, or luminous rings round the Moon's disk, are comparatively common, and can be beautiful. They are due not to dust, but to moonlight shining upon ice crystals in the upper air, about 20,000 feet above the ground. These crystals make up the type of cloud known as cirrostratus. If the cloud is lower and denser, the Moon merely looks watery. Both watery moons and haloes are said to be forerunners of rain, and this is often true, because cirrostratus cloud is itself a common sign of approaching bad weather. Paraselenæ, or 'mock moons'—

brilliant images of the Moon some way from the true disk—are also due to ice crystals, but are very rare. I have yet to see one.

When the Moon shines upon water droplets in the atmosphere it may produce a rainbow in the same way as the Sun, but lunar rainbows are relatively faint and rare, as well as being less brightly coloured, simply because the intensity of moonlight is so much less than that of sunlight; as I have already said, it would take more than half a million full moons to provide as much light as the midday Sun. Yet now and then a striking lunar rainbow is seen. Flying at about 2,000 feet above Scotland during a war-time exercise on 28 March, 1945, I was particularly lucky; most of the rainbow circle could be seen, and even some delicate, fugitive hues, giving a lovely effect. As I was the navigator of the aircraft I had little time to study the rainbow, but I am glad to have seen it.

The celebrated moon illusion, which has been the subject of a good deal of research, is not an atmospheric phenomenon at all, and in fact it still remains something of a puzzle. For some reason or other, the full moon seems to look larger when close to the horizon than when it is high in the sky. Indeed, the casual observer is apt to say that it looks twice the size, whatever the state of the sky—particularly at the time of Harvest Moon. Actually, the Moon is slightly more distant when low down than when high up. The observer is brought toward the Moon as the Earth turns, so that the high Moon is a little the larger, though the difference amounts to less than two per cent. Why, then, should the low Moon look the bigger?

It is not really so, as measurements prove, so that some trick of the eye or the brain is responsible. (In passing, the Moon never looks so large as most people imagine; try covering it with a small coin held at arm's length, and you will see what I mean.)

The illusion has been known for many centuries. So far as I know, the first man who tried to explain it was Ptolemy, around A.D. 150, and his theory is certainly better than many of those put forward in modern times, though it is not likely to be the complete answer. Ptolemy pointed out that the low-down Moon is seen against a foreground of 'filled space' (trees, houses, etc.), and so there are nearby objects to act as comparisons; when the Moon is well above the horizon there are no comparisons, and we look at it across 'empty space'. When low, then, the Moon will seem to be more remote than when it is high; and if the images seen in the eye are of equal size, the disk which is the further away will seem to be the larger.

Yet will the Moon really seem more distant when we look at it behind 'filled space'? One man who disagreed was George Berkeley, who in 1709 published a completely different theory. First he tried to dispose of Ptolemy's idea, by looking at the rising full moon through a tube and thereby cutting out any foreground; he claimed that the illusion was still obvious. He went on to suggest that because the low Moon is shining through a relatively thick layer of the Earth's air, it appears fainter than the high Moon (because more of the moonlight is absorbed), and it is this which makes the low Moon seem both more remote and oversized. Berkeley's theory was supported in 1973 by Professor E. J. Furlong of Dublin University, though in modified form. Furlong held that a change in the Moon's colour, again due to its being seen through a thicker layer of atmosphere, is an important factor.

Some years ago E. G. Boring, at Harvard, carried out experiments which indicated that the illusion is due to the mechanism of the human eye; the unconscious effort of raising the eye to look at a high-up object causes the Moon to look smaller than it really is. Boring's work was followed up in 1959 by H. Leibowitz and T. Hartman, at Wisconsin, who experimented with disks seen at eye-level and overhead. They concluded that the illusion is due to the fact that we have more visual experience with objects in the horizontal plane than in the vertical. They also added that children were more affected than adults.

The next major paper on the subject, by L. Kaufman and I. Rock in 1962, was essentially a return to Ptolemy's theory. Other writers claimed that one-eyed observers were not conscious of the illusion, and there was also a suggestion that dustiness in the atmosphere near the horizon blurs the edges of the Moon and enlarges the disk.

It all seemed very uncertain, and in 1974 I presented a television programme about it. With me was Professor Richard Gregory, who is an acknowledged expert upon illusions in general (as well as on many other subjects!). The first thing we tried to do was to test the Berkeley-Furlong theory. We fixed up a white disk, and illuminated it from behind; then we observed it from a distance, dimming it and changing its colour by means of filters put in front of it. There was no perceptible difference in apparent size, so we turned to an experiment entirely our own —though I must stress that Professor Gregory designed it, and that I was merely the operator.

We went out on to the Selsey coast at the dead of night, armed with a system of movable mirrors. The full moon was high up, and the sky was clear. We could produce an artificial moon whose size could be altered by means of an iris, and by swinging the mirror we could alter the apparent altitude of the image of the real Moon. The idea was to test not the real sizes of the disks, but their apparent relative sizes, which is not the same thing. While the senior experimenter operated the mirrors and lights, I stood well back and compared the two disks, estimating them as accurately as I could and using a special measuring device. First the real Moon was 'brought down' until it was side by side with the artificial one, and the iris was adjusted until, to me, the two disks were equal. Then the real Moon was raised—and it seemed to shrink; only when the artificial image had been reduced by about 10 per cent. did the two seem equal again.

We had also persuaded a member of our team to cover up one eye for several hours beforehand, and found that the illusion was still there. To test the behaviour of my eyes when looking at objects from unfamiliar angles, I observed both the low Moon and the high Moon while standing on my head, thereby causing onlookers to class me as insane. Again there was no difference in the illusion. We finally concluded that Ptolemy had been on the right track, but that foreground, eye and brain were all involved. Since the illusion is essentially a physiological one, and is not due to the Moon itself, further discussion of it here would, I fear, be out of place.

'Lunar' and 'lunacy' have long been associated, and it is often said that people who are mentally unstable are at their very worst when the Moon is full. On this point, medical experts seem to be divided. I have consulted at least twenty mental specialists, and the general consensus of opinion seems to be that there is no connection, but I must admit that there were several dissentients. (As I have absorbed more concentrated moonlight into my eyes than most people, I rather hope that there is no truth in the idea.) Of course, certain small creatures regulate their habits according to the Moon, but all these creatures are aquatic, and it seems reasonable to suppose that the tides are responsible rather than the Moon itself. Land animals and plants are, apparently, not affected. Fifty years ago many farmers still believed that it was unwise to sow crops at full moon, but controlled scientific experiments have given negative results. It is safe to say that even if there is any relationship between crops and the Moon's phases it is of no practical importance.

It would be pointless to discuss all the various superstitions which have grown up around the Moon, such as the old story that it is unlucky to see the new moon through glass; and I really cannot bring myself to waste any time upon astrology, the so-called 'science' which held back the progress of true astronomy for so long. According to astrologers, the positions of the Sun, Moon and planets in the sky affect human characters and destinies, and of course the Moon is regarded as highly potent. Generally speaking, this sort of thing is harmless, but an astrologer with genuine powers is about as common as a great auk.

It seems, then, that the Moon's main influence upon us concerns the movement of the Earth and the regulation of the tides. Once we try to take matters any further, we are treading upon very uncertain ground indeed.

Finally, there are the mysterious objects known as tektites (Fig. 21), which have caused a tremendous amount of discussion over the years. This is not really a digression, because there have been suggestions that tektites come from the Moon and have been hurled across space to the Earth. The idea seems to have been originally proposed by a Dutch scientist, R. D. M. Verbeek, as long ago as 1897, and it still crops up now and then.

Tektites are small, dark, glassy-looking objects which are found only in certain regions of the Earth. Most of them are less than an inch long, and even the largest known example is smaller than a hen's egg. They contain about 80 per cent. of silica, and are easy to recognize, since they are unlike any other natural objects so far discovered.

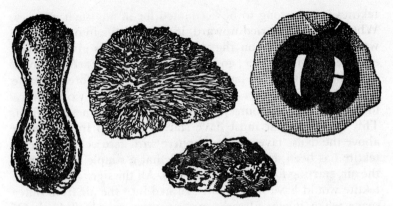

Fig. 21. Tektites

Their distribution is remarkable. The widest tektite field is in Australia, covering much of the southern part of the continent; at least two more fields lie in the East Indies area, and there are a few others in Asia and America. The main European field lies in Czechoslovakia. No tektite has ever been found in Britain.

Originally, it was thought that tektites must be of terrestrial origin. Even a casual glance was enough to show that at one stage in their careers they had been very strongly heated, and the obvious answer was that they had been hurled out of active volcanoes. The main objection to this theory is that tektites are found in areas well away from volcanoes of any sort, and on the geological scale they are not very old, though of course their ages amount to millions of years. (The Australian tektites are the youngest.) Also, some tektites at least seem to have been heated not once, but twice. The first heating was apparently thorough, but the second was less violent, affecting only the layers near the surface.

This might be explained on the volcano theory—or so it was thought. Suppose that a tektite were heated inside a volcano, and then shot out at a speed great enough to carry it beyond the densest part of the Earth's atmosphere? As it fell back toward the ground, it would be re-heated by friction against the air-particles in the same way as a meteor.

However, D. Chapman, in California, carried out a full study of the problem, and showed that this explanation, simple though it sounds, does not fit the facts. The second heating of

71

tektites is too strong to be accounted for in any such manner. When a body is hurled upward, it may behave in one of three ways. Depending upon the velocity which it is given, it will either escape into space, go into orbit round the Earth, or come to a stop and then fall back to the ground. Therefore, a tektite which began its career inside an Earth volcano would have to follow the third course, as otherwise it would never return. This means that it must have started its drop from a point above the dense layers of atmosphere with zero velocity. Yet a tektite has been so intensely heated that a simple fall through the air, starting from rest, will not do. All the signs are that the tektite would have had to have rushed into the air from outer space when it was already moving very quickly indeed. Of course, the greater the initial velocity, the greater the heating-up by friction against the air.

Efforts to explain tektites by normal geological processes have been uniformly unsuccessful. (In less serious vein, there have been various peculiar theories proposed from time to time; it has been suggested, for instance, that our remote ancestors—possibly the Ancient Atlanteans—were remarkably good at glass-making, while a Soviet writer named Agrest stoutly maintains that tektites mark the traces left by nuclear-powered space-ships which have visited the Earth from another planet!) Also, tektites are aerodynamically shaped, which again points to a descent through the atmosphere.

All this is very significant. Tektites are quite unlike ordinary meteorites; they have had their two heatings, the first one so intense that it could have taken place in a volcano. And if Earth volcanoes cannot be responsible—well, there must once have been plenty of active volcanoes on the Moon, which brings us back to Verbeek.

Plausible-sounding ideas were put forward. According to one of them, tektites were born inside lunar volcanoes, and were hurled out so furiously that they escaped from the Moon altogether, cooling down when they reached what is often called 'outer space'. Then they came into the Earth's atmosphere, and were re-heated during their fall to the ground. Each tektite field would be related to some specific lunar eruption, but few eruptions would produce material moving in a suitable manner to hit the Earth, which explains why tektite fields are rare. Alternatively there is a proposal by H. H. Nininger, the American expert on meteoritics, that tektites were shot away from the Moon when large meteorites struck the surface.

The 'lunar hypothesis' was popular some years ago, but we have new information now. If tektites originated on the Moon, we would expect to find at least some on the surface rocks now, but we do not. Among all the samples brought home by the Apollo astronauts and the Russian automatic probes, not a trace of tektite material has been found, and it does not seem likely that there are tektites anywhere on the Moon—though it is true that as yet we have samples only from a few localized areas.

The mystery remains. We simply do not know whether tektites are of terrestrial or extra-terrestrial origin, but I must leave the problem there, if only because we can be reasonably sure that they do not come from the Moon. It was only with the return of Apollo 11, in July 1969, that Earth scientists could analyze their first lunar rocks. Men were forced to go to the Moon or send automatic vehicles there to collect specimens of material; Nature had not been kind enough to do the work for us.

Chapter Seven

OBSERVERS OF THE MOON

THE FIRST KNOWN MAP of the Moon was drawn around the year 1600. It was the work of William Gilbert, physician to Queen Elizabeth I, who is best remembered today as the pioneer investigator of magnetic phenomena. Gilbert's map was not actually published until 1651, but it must have been completed before the end of 1603, for the excellent reason that this was the year in which Gilbert died. The main dark regions are shown in fairly recognizable form; for instance, the patch which he calls the 'Regio Magna Orientalis' is identifiable with the vast plain known to us as the Mare Imbrium.

There are two points of special interest about Gilbert's map (Fig. 22), quite apart from its being the first. He regarded the dark areas as land and the bright regions as seas—the reverse of the general view at that time. More importantly, Gilbert drew his chart with the naked eye. Telescopes did not come upon the astronomical until the end of the first decade of the seventeenth century.

I have always been surprised that the telescope was not invented much sooner. After all, spectacle lenses to help people with poor eyesight were certainly in use by 1300, and by Gilbert's time spectacles had begun to look rather like ours. So far as we can tell, the inventor of the telescope was a Dutch spectacle-maker, Lippershey, who produced one in 1608. But this was after Gilbert's death, and it seems that the Gilbert map was the one and only naked-eye chart of the Moon ever drawn in pre-telescopic times.

The other thing which puzzles me is why—with respect to Gilbert—his map was not better than it proved to be. Some years ago I carried out an experiment in which I asked various people to look at the full moon and draw what they could see. The results were surprisingly good. I admit, of course, that most of the volunteers had seen photographs of the Moon, so that unconscious prejudice was bound to have an effect, but I did my best to select normal-sighted people whose interest in astronomy was nil.

74

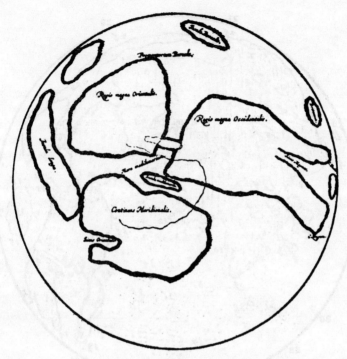

Fig. 22. William Gilbert's naked-eye Moon map

At any rate, Gilbert had made a start (even if he did not publish it). The real study of the Moon could begin as soon as telescopes were available. Of the early observers, Galileo was by far the greatest, but he was not the first. This honour seems to go to Thomas Harriott, one-time tutor to Sir Walter Raleigh, who was an excellent mathematician, and who evidently managed to lay his hands on one of the telescopes built on the mainland of Europe. I feel able to make a comment here, because I played a rôle, albeit a very minor one, in the first publication of Harriott's lunar map (see Fig. 23).

In 1965 I had a letter from Dr. E. Strout, of the Institute of the History of Science in the U.S.S.R., enclosing a paper he had written; he had even located the map, which had apparently been drawn in July 1609. I was then (as now) Director of the Lunar Section of the British Astronomical Association, and I was able to arrange the prompt publication of the Strout paper,

75

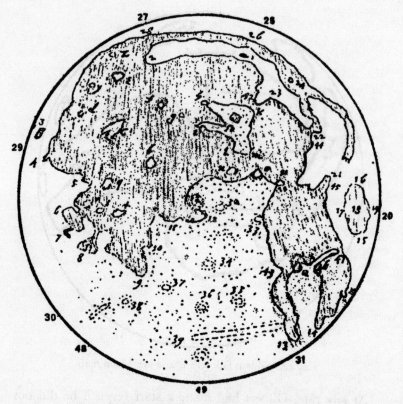

Fig. 23. Thomas Harriott's telescopic Moon map

plus the Harriott map, in the Association's *Journal*. (Volume 75, page 102.)

Since Galileo did not begin his telescopic work until the winter of 1609-10, Harriott anticipated him—assuming, of course, that his map really does date from July 1609, which is not absolutely certain. In any case, the chart contains many features which are clearly recognizable, and in view of the weakness and poor definition of telescopes at that time it was a remarkably good effort—much better, candidly, than anything which Galileo produced. Unfortunately Harriott never followed it up. He died in 1621.

One of Harriott's correspondents was Sir William Lower, who lived in the Welsh village of Treventy. Lower is a shadowy, elusive figure. He was certainly expelled from Oxford University in 1591, following a celebration which was presumably anything but teetotal; from 1601 to 1611 he sat in Parliament, first as member for Bodmin and then for Lostwithiel; he was knighted in 1603, and after his marriage to one Penelope Perrot, daughter and heiress of Thomas Perrot of Treventy, he retired to the seclusion of Wales to spend much of his time in studying astronomy. Very little is known about his work, but he did leave a graphic description of the full moon as resembling a tart that his cook had made—'there some bright stuffe, there some dark, and so confusedlie all over'. This does not sound ultra-scientific, and I have been unable to discover any observations by Lower which were of true value; but at least he looked at the Moon, and we may suppose that he went on doing so until he died in 1615. I would not suggest that Lower is important in the story of selenography or lunar study, and I mention him here only because most astronomical historians ignore him completely.

So let us turn to Galileo, one of the most famous of all scientists. I do not propose to say much about his life, because it is so well-known, and in any case I have attempted a comprehensive account elsewhere.* It is enough to say that from 1609 onward he made telescopes of his own, and used them to make a whole series of dramatic discoveries. Naturally he looked at the Moon, and he left a good description of it, as follows:

. . . I distinguish two parts of it, which I call respectively the brighter and the darker. The brighter seems to surround and pervade the whole hemisphere; but the darker part, like a sort of cloud, discolours the Moon's surface and makes it appear covered with spots. Now these spots, as they are somewhat dark and of considerable size, are plain to everyone, and every age has seen them, wherefore I will call them *great* or *ancient* spots, to distinguish them from other spots, smaller in size, but so thickly scattered that they sprinkle the whole surface of the Moon, but especially the brighter portion of it. (These features are, needless to say, the craters.) These spots have never been observed by anyone before me; and from my observations of them, often repeated, I have been led to the opinion which I have expressed, namely, that I feel sure that

* In my book *Watchers of the Stars* (London, 1974).

77

the surface of the Moon is not perfectly smooth, free from inequalities and exactly spherical, . . . but that, on the contrary, it is full of inequalities, uneven, full of hollows and protuberances, just like the surface of the Earth itself, which is varied everywhere by lofty mountains and deep valleys.

This sounds 'modern' enough, and Galileo went on to make an effort to measure the heights of some of the lunar mountains. His results were obviously rough, but they were better than nothing at all, and were at least of the right order of magnitude. Galileo found that the lunar mountains are much loftier, relatively, than those of the Earth; the Moon is a rugged world.

Galileo made a number of lunar drawings, and on some of them it is possible to identify various features, though the accuracy is low (Fig. 24). Unlike Harriott, Galileo never compiled a proper map, and it seems that after those early days he paid little further attention to the Moon's surface; he was busy with greater problems—and, of course, with his battles against the Church, which make fascinating reading but which do not concern us here. Remember, too, that Galileo's telescopes were very feeble. Even the most powerful of them magnified no more than thirty times.

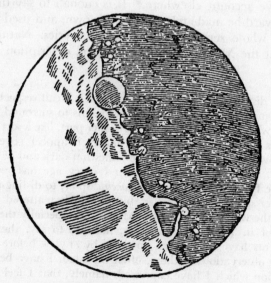

Fig. 24. Galileo's drawing of the Moon

Galileo made one more comment about the Moon in 1641, after the infamous trial at the hands of the Inquisition, and when he was old, blind, and living under strict Church supervision in his lonely villa. A scientist named Fortunio Liceti had claimed that the faint luminosity of the night side of the Moon, seen during the crescent stage, was due to the diffusion of sunlight in a lunar atmosphere. Galileo argued, quite correctly, that Leonardo da Vinci had been right in saying that it must be due to light reflected on to the Moon from the Earth.

The first stage of lunar telescopic research ended with Galileo. For the first time the mountains and craters could be clearly seen; it was still generally believed that the dark areas were watery, and there seemed no obvious reason to doubt that there could also be life. One man who certainly thought so was Johannes Kepler, who corresponded with Galileo and who drew up the famous Laws of Planetary Motion which gave the death-blow to the theory that the Sun goes round the Earth. Kepler went so far as to write a science-fiction story in which he gave a vivid description of the lunar life-forms—some of them serpent-like, others aquatic, and all of them covered with thick fur! Kepler's story, the *Somnium*, was published posthumously in 1634.

The next major advance came in 1647, and was due to Hevelius (more properly, Hewelcke), a city councillor of Danzig in Poland, now known as Gdańsk. Hevelius built an observatory on the roof of his house, and equipped it with the best instruments available at the time. His telescopes were strange and awkward; they had small object-glasses and very long focal lengths, and were incredibly cumbersome. (In some telescopes the object-glass had to be entirely separate, and fixed to a mast. A French astronomer named Auzout once designed a telescope 600 feet long, though apparently it was never built.) Yet Hevelius was a patient, skilful observer, and his map (Fig. 25), just under a foot in diameter, showed many identifiable features. He also made height-measurements of some of the lunar peaks which were better than Galileo's.

Hevelius gave considerable thought to the best method of naming lunar features. He finally decided to give them terrestrial names, and this was the system followed on his map. For

Fig. 25. Hevelius' Lunar map

instance, the crater now called Copernicus was his 'Etna', while our Plato was 'the Greater Black Lake'. It was a reasonable idea, but it did not work well, and less than half a dozen of Hevelius' lunar names are still in use.

Copies of the map exist, but the original copper-plate of it is no longer to be found. After Hevelius' death it was apparently melted down and made into a teapot.

Next came Giovanni Battista Riccioli, an Italian Jesuit who published a lunar map in 1651. It was about the same size as Hevelius', and was based largely upon the observations made by Riccioli's pupil, Grimaldi. The map itself was fairly good, but it is remembered chiefly for its system of nomenclature. Riccioli gave the dark plains romantic names such as the Mare Imbrium (Sea of Showers), Mare Tranquillitatis (Sea of Tranquillity) and Sinus Iridum (Bay of Rainbows); the craters were named after famous personalities—usually, though not always, astronomers. The system quickly replaced that of Hevelius, and nearly all the names he gave (more than 200 altogether) are still in use. Later selenographers have added to the list, and the whole scheme is so deeply rooted that it will certainly never be altered now.

Riccioli was a man of firm opinions, some of which were completely wrong. He could never bring himself to believe that the Earth is in orbit round the Sun, and this meant that he had little patience with Copernicus, who had revived the Sun-centred or heliocentric system over a hundred years earlier. Alas, Copernicus had to have a crater; but Riccioli, to quote his own words, 'flung Copernicus into the Ocean of Storms', which is why the formation honouring the great Polish scientist is to be found in the midst of the vast Oceanus Procellarum. At least the crater is large and important, but Galileo was not so well treated, and was allotted a very obscure crater toward the edge of the Oceanus. On the other hand, Ptolemy was given a magnificent walled plain close to the middle of the Moon's disk, and to Tycho Brahe, the eccentric Dane, went the most prominent crater on the Moon; it lies in the southern uplands, and is the centre of the ray-system which dominates the whole lunar scene near Full. Riccioli was not modest. He and Grimaldi are represented by two huge, dark-floored walled plains which are recognizable at any time when they are in sunlight.

One trouble about all this was that most of the chief craters were used up, so that later students of the Moon had to be content with second best. Sir Isaac Newton, who was a small boy when Riccioli drew his map, is tucked away near the Moon's south pole; Johann Mädler, about whom I shall have much to say shortly, has an insignificant crater on the Mare Nectaris or Sea of Nectar, and so on. There are also some rather unexpected names. Julius Cæsar is there, presumably because he was responsible for the reform of the calendar; so are Alexander the Great and his friend Nearch; there are even a couple of Olympians, Atlas and Hercules. One crater has been given the rather startling name of Hell. This does not, however, indicate exceptional depth; it commemorates Maximilian Hell, a Hungarian astronomer of the eighteenth century. Also on the Moon we find Barrow (Isaac Barrow, Newton's tutor) and Birmingham (John Birmingham, an Irish observer who died in 1884).

At one stage, not so long ago, the whole scheme of lunar nomenclature began to get out of hand. Finally, the International Astronomical Union, the controlling body of world astronomy, standardized matters and the situation today is much more satisfactory. Of course, it has also been necessary to name the features on the Moon's far side, never visible from Earth; but that is running ahead of our story.

Passing over a hundred years, and pausing only to mention the 21¼-inch map of the Moon drawn by Giovanni Cassini in 1680, we come to Tobias Mayer, a German astronomer whose lunar chart was published in 1775, thirteen years after his death. Mayer introduced a system of co-ordinates (the equivalents of latitude and longitude on the Moon) and produced an excellent 8-inch map which remained the best for over half a century. Unlike Hevelius, Riccioli and most of his predecessors he drew the Moon with south at the top, which is how telescopic observers actually see it—provided, of course, that they live in the northern hemisphere of the Earth.

Most astronomical telescopes give a reversed or upside-down image, which in general does not matter in the least. To make the image erect means putting an extra lens system into the telescope—and since this must involve some loss of light, there is no point in adding it, unless the telescope is also to be used for mundane observations such as looking at ships out at sea (for which, let it be added, a telescope designed for astronomy is quite unsuited). It was therefore logical to print all telescopic maps with south at the top and north at the bottom; conventionally 'west' on the Moon meant to the left-hand side of the telescopic image, as shown in Fig. 30 (top left). Only in very modern times has there been a change.

Soon after the appearance of Mayer's map, Johann Hieronymus Schröter founded a private observatory at Lilienthal, near Bremen in Germany, and embarked upon a research programme which was really the beginning of modern-type selenography. He was never a professional astronomer. For much of his career he was chief magistrate of Lilienthal, with ample means and leisure to carry on his hobby, and he acquired several reflecting telescopes, two of which were made by no less a person than William Herschel. The largest instrument had a 19-inch mirror, and was the work of Schräder of Kiel. For thirty years he worked patiently away, drawing, measuring and charting (Fig. 26). To a great extent he was breaking new ground, and it was he who, for instance, first studied the crack-like features which we now call clefts or rills. He did not actually discover them (this honour must go to the great Dutch scientist Christiaan Huygens, a century earlier), but Schröter was the first to chart them in detail.

Schröter has been much maligned. It has been claimed that all his work was inaccurate, and that even his telescopes were of poor quality. I can only say that I disagree most strongly. It is true that he was not a good draughtsman; his drawings are clumsy and schematic, and the 19-inch Schräder reflector may well have been imperfect. Also, some of Schröter's ideas were very wide of the mark. He believed that he had found important structural changes on the Moon's surface, he considered that there must be a fairly dense atmosphere, and he was quite prepared to accept the idea of intelligent life there. On the other hand he was a completely honest observer, who never drew anything unless he could be quite certain in his own mind that he had seen it, while his measurements of mountain-heights were better than any previously made. As for his telescopes—well, whatever may be said about the 19-inch, there can be no doubt about the high quality of the two made

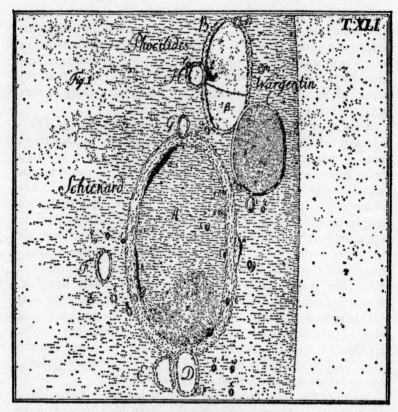

Fig. 26. Lunar drawing by Schröter

by Herschel.* The more I study Schröter's work, the greater my admiration for it.

Unfortunately, despite his hundreds of drawings of various parts of the Moon, he never made a complete map, and his career came to a sad and abrupt end. In 1813, when he was approaching the age of seventy, the French, under Vandamme, occupied Bremen; Lilienthal fell into their hands, and Schröter's observatory was burned to the ground, along with

---

* As an interesting historical aside; so far as we know, the only telescopes made by Herschel which were ever used for serious work were his own and Schröter's. Yet during his lifetime Herschel built, and distributed, many dozens of high-quality instruments.

all his notes, manuscripts and unpublished observations. Even the telescopes were plundered, because they were brass-tubed, and the French soldiers thought that they must be gold. The loss could never be made good, and the old astronomer had no time to begin again; he died three years later.

The mantle of Schröter fell upon three of his countrymen: Lohrmann, Beer and Mädler. All were clever draughtsmen as well as being good observers, and between them they explored every square mile of the Moon's visible surface—but they had Schröter's work to use as a basis. The credit for founding the science of precise lunar observation should go to the Lilienthal amateur above all others.

Wilhelm Lohrmann, a Dresden land surveyor, set out to compile a really large, accurate lunar map. The first sections were published in 1824, and were amazingly good. Unfortunately he had completed only three more sections when his eyesight failed him, and he had to give up. He died in 1840.

Next came the superb work by a Berlin banker, Wilhelm Beer, and his friend Johann Mädler. They used the 3¾-inch refracting telescope at Beer's home, and studied the Moon intensively for more than ten years, finally producing a map which was the basis for all later studies right up to the start of the Space Age. They followed it up with a book, *Der Mond*, which contains a full description of the whole of the visible surface. *Der Mond* appeared in 1838, and made a tremendous impact upon the astronomical world. For some strange reason it has never been translated into English. (Had I been able to read German, I would have put this right years ago.)

Oddly enough Mädler, who did most of the mapping, used the 3¾-inch telescope for almost all his lunar work. It is true that, inch for inch, a refractor is more effective than a reflector; even so, the difference between Mädler's telescope and Schröter's instruments is remarkable. No doubt Mädler's gave a sharper image, but it must have been inferior in sheer light-grasp.

Der Mond not only revolutionized selenography, but also actually held it back to some extent! Schröter had believed the Moon to be a living, changing world, but Beer and Mädler went to the other extreme, and considered that it must be

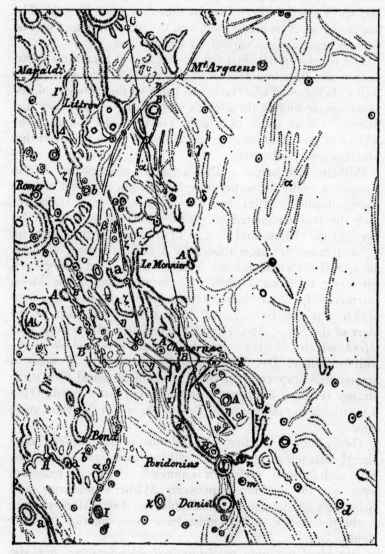

Fig. 27. A map from Edmund Neison's *The Moon*, published 1876

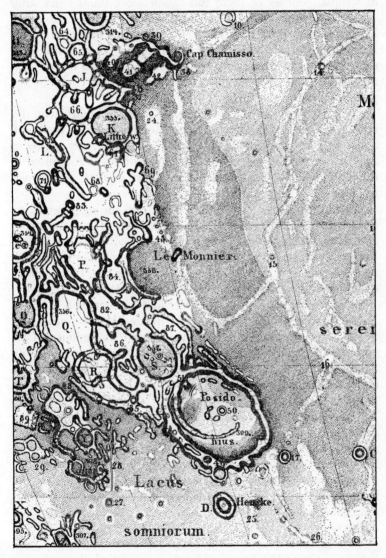

Fig. 28. The same area as in Fig. 27 opposite, taken from Dr. Julius
Schmidt's *Mondcharte*, 1878

absolutely dead. Their opinions naturally carried a great deal of weight. Neither of them did much more lunar work after 1840, when Mädler left Berlin to become director of the new Dorpat Observatory in Estonia (then, as now, occupied by the Russians), and nobody else seemed really inclined to follow in their footsteps. If their map were the last word on the subject, and if the Moon were a changeless world, what could be the point of observing it further?

Luckily there was one astronomer, Julius Schmidt, who did not agree. He began observing the Moon during his boyhood, and continued to do so until his death in 1884. After acting as assistant at various German observatories he was appointed to the directorship of the Athens Observatory in 1858, and it was in Greece that most of his best work was done.

Schmidt not only revised and completed the map begun by Lohrmann, but also produced one of his own. It was 74 inches across, and stands up well when compared with modern charts. Yet before this map appeared—in 1878—much had happened, and Schmidt was chiefly responsible. It was the 'Linné affair' which reawakened popular interest in the Moon.

At various times Lohrmann, Beer and Mädler, and Schmidt himself, had recorded a deep crater in the Mare Serenitatis or Sea of Serenity. Mädler had named it Linné in honour of Carl von Linné (Linnæus), the Swedish botanist. Then, in 1866, Schmidt announced that the crater was no longer there. It had vanished from the Moon, or at least altered in appearance beyond all recognition.

This was startling, to put it mildly. Could the Moon be less dead than Mädler had thought? It was a revolutionary idea, and yet, coming from an observer with Schmidt's reputation, it could not be disregarded. Amateurs and professionals alike were intrigued, and once more telescopes were pointed at the lunar surface. The number of papers and discussions about this minor feature of the Moon became astronomical in every sense of the word.

To anticipate Chapter 15, I will say here that in my view there was no real change in Linné; even skilled observers can be deceived. Yet the whole affair was beneficial to the cause of selenography, and ever since 1866 the Moon has been kept under close watch.

Up to that time the Germans had taken the lead, but after the Linné alarm others joined in. The first important British book on the subject was the work of Edmund Nevill, who wrote under the name of Neison. *The Moon*, published in 1876, contained a map based on Mädler's, together with a description of every named formation. It was of tremendous value; even today copies of it can be picked up occasionally, and even when I began looking at the Moon, around 1930, it was still regarded as a 'must' for every lunar observer.

Neison himself provides a link between the past and the near-present. He was only twenty-five when he wrote his book, and when I first read it I naturally thought that he must have been dead for many years. In fact he was not; he was living at Eastbourne, only thirty miles from my old home at East Grinstead, and I could well have met him, though to my great regret I never did so. He died in 1938. He did very little lunar work after 1882, when he went to South Africa to become director of the new observatory in Natal—which was unfortunately closed down in 1912 because of lack of funds.

At about the time that Neison's book was published, a new society devoted entirely to studies of the Moon was formed in England. It was called the Selenographical Society, and for ten years or so it was very active. In 1883, following the death of its president (W. R. Birt) and the resignation of its secretary (Neison) it was disbanded, but seven years later the newly-founded British Astronomical Association set up a Lunar Section and carried on the work.

From the beginning the B.A.A. has been made up chiefly of amateurs; and since the Moon remained more or less of an amateur province until the start of practical space research, there was no shortage of observers. Ever since 1890 the Lunar Section has kept up its work, and it still does, as will become apparent when we come to consider what I have christened T.L.P. or Transient Lunar Phenomena. Eleven full-length memoirs have been published, as well as dozens of papers and reports scattered through the Association's regular journal. Thomas Gwyn Elger, first director of the Lunar Section, published a book on his own account in 1895, illustrated by an outline map which was very useful indeed.

Meanwhile, the camera had started to make its presence felt.

The very word 'photography' was first coined by an astronomer —Sir John Herschel, son of Sir William—and its power in lunar work was seen as early as 1839, when François Arago described the invention in his speech to the Chamber of Deputies in Paris. Arago's words are worth quoting: 'We may hope to make photographic maps of our satellite, which means that we will carry out one of the most lengthy, most exacting, and most delicate tasks of astronomy in a few minutes.' In fact things were not so simple as Arago had expected, and it is probably true to say that lunar mapping was completed only in 1967, with the photographs sent back by the Orbiter rocket probes; but it was not long before the camera started to replace the eye in many ways.

On 23 March, 1840 J. W. Draper, of New York, made a Daguerreotype of the Moon, using a long-focus 5-inch reflector. The image was a mere inch in diameter, and the exposure-time needed was as long as twenty minutes, but various light and dark areas were recorded. Ten years later J. A. Whipple, at Harvard, obtained a good series of Daguerreotypes showing the Moon at different phases, and some of these pictures were capable of being enlarged to a scale of five inches to the Moon's diameter. (I remember including some of them when I took part in arranging the Moon display at the Festival of Britain in 1951. They had also been shown at the Great Exhibition of 1851, causing a great deal of interest.)

The development of the wet collodion process helped matters considerably, and in 1852 some good results were obtained by an English amateur, Warren de la Rue. Later, using dry plates, de la Rue in England and Rutherfurd in the U.S.A. were able to take photographs which were of definite astronomical value. Others, too, began to join in, and after 1890 lunar photography became all-important.

During the last decade of the eighteenth century and the first of the nineteenth, true photographic atlases began to appear: one from Paris, one from Belgium, one from the Lick Observatory in America, and so on. The two main Paris observers, Loewy and Puiseux, produced an excellent series of plates covering the entire visible face of the Moon, and subsequently all lunar mapping was photographically based. S. A. Saunder, an English amateur who was by profession a schoolmaster, drew up a list of the positions of lunar features which was of tremendous value; and in 1904 W. H. Pickering, one of the few professional astronomers to show a real interest in the Moon, issued an atlas in which he showed each region under five different conditions of illumination.

Yet despite these and other atlases, there followed another period of relative neglect. Professionals, in general, had no time to spare for lunar charting, and tended to regard the Moon as rather dull and parochial. The new giant reflectors were seldom used for selenography. Some trial exposures were made around 1919 with the new 100-inch reflector at Mount Wilson, in California, but on the whole the Moon was still regarded as an amateur field.

Nobody can deny that the amateurs rose nobly to the occasion. In 1930 Walter Goodacre, who had succeeded Elger as Director of the B.A.A. Lunar Section, published a book containing a detailed map as well as a full account of each named formation. Even larger was the map drawn by H. P. Wilkins (Fig. 29), who directed the Lunar Section between 1946 and 1956. His final chart, to a scale of 100 inches to the Moon's diameter, was published in 1959. In that year he retired from his official post as a civil servant in the Ministry of Supply so that he could spend the rest of his life in revising his maps, and it was a tragedy that he should die of a heart attack in the following January. He was, I suppose, the last of the great 'visual mappers' of the Moon.

The whole situation was changing—and changing fast. As I have said, professional interest in the Moon had largely lapsed between the two world wars. It is true that the International Astronomical Union issued a map in 1935, drawn by W. H. Wesley and M. A. Blagg and based upon the earlier measures by Saunder and his colleague Franz, but it was hardly a success; it was not clear enough to be really useful, and it was defective near the Moon's limb. But when the rockets began to fly, and it became clear that the Moon would be reached much sooner than most people had thought possible, a really detailed and accurate map was needed. Otherwise, it would be unthinkable to dispatch astronauts to the Moon.

The obvious solution was a complete photographic chart, similar to Pickering's atlas or the Paris atlas, but much larger

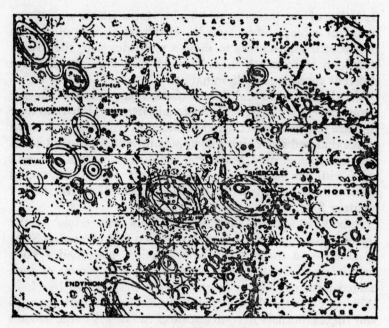

Fig. 29. Part of Wilkins' Moon map

and sharper. This was attempted by G. P. Kuiper and his colleagues at the Yerkes Observatory in America, and appeared in 1959. It was based on the best photographs available—including those taken by Gilbert Fielder at the high-altitude Pic du Midi Observatory in the French Pyrenees—and it represented a really major contribution. Yet even so, it did not meet all the requirements. Over-enlargement led to blurring of many of the features, except near the centre of the disk, and the limb regions, alternately carried in and out of view by libration, were poorly shown.

This was not the fault of the observers, or of their equipment. The trouble lay partly with the Earth's inconvenient, unsteady atmosphere and partly with the Moon itself—since all the features near the limb are badly foreshortened. Absolutely nothing could be done about this short of going into space.

It is untrue to say that no good atlases have been compiled by Earth-based observers since then. There is, for instance, the Japanese atlas, produced at the Kwasan Observatory by

S. Miyamoto and his colleagues. I must make special mention of the wonderfully good photographic atlas by the English amateur Commander H. R. Hatfield,* which—like Pickering's, so long ago—shows the whole of the visible disk under several illuminations, and which I regard as essential for any practical observer. But so far as lunar travel was concerned, no map depending upon observations made from Earth could be detailed enough.

Space research methods took over, and caused a minor upheaval in accepted procedure. Some astronomers wanted to go on printing maps and photographs with south at the top; rocket men wanted to have north at the top. Finally, there was a debate in the International Astronomical Union, ending in a vote. I spoke on behalf of the defence, but I was badly defeated; the 'north-toppers' won the day, and even now long-standing observers such as myself have to turn official maps upside-down before being able to make sense out of them! At the same time east and west were reversed, so that today west is to the left with north at the top (see Fig. 30). One amusing result of this was that the Mare Orientale or Eastern Sea is now in the *west* of the Moon. I cannot help smiling wryly, since Wilkins and I named this particular feature when we discovered it just after the end of the war . . .

The first successful Moon rocket was Lunik 1 (or Luna 1), sent up by the Russians in January 1959. Eight months later Luna 2 crash-landed on the surface, and then, in October 1959, came the first photographic probe—Luna 3, which went round the Moon and sent back pictures of the previously unexplored far side. The next step was delayed until the middle of 1964, when the Americans had their first major success; Ranger 7 sent back more than 4,000 photographs of part of the Mare Nubium before destroying itself on impact. Two more Rangers followed. But it was with the five Orbiters that lunar mapping was really completed.

I will say more about the space-research programme in Chapter 12. Meantime, let us confine our attention to the results from 1966, when Orbiter 1 was launched, to the summer of 1967, when the fifth and final Orbiter went on its way. The

* *The Amateur Astronomer's Photographic Lunar Atlas* (Lutterworth Press, London, 1969).

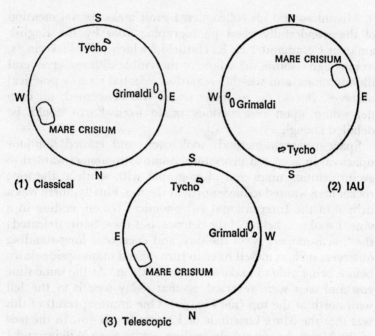

Fig. 30. Various systems of orientation. (1) Classical—south at the top, Mare Crisium to the west. (2) IAU: Mare Crisium to the east. (3) The telescopic view as used in this book. I have retained south at the top, but have accepted the IAU ruling of putting Mare Crisium *east* and Grimaldi *west*

probes put into paths round the Moon, and sent back so many photographs that even today there are thousands waiting to be properly examined and analyzed. Virtually the whole of the lunar surface is covered, and of course we can get rid of the awkward foreshortening effects which make the limb regions so hard to map from Earth. Look, for instance, at two views of the walled plain Wilhelm Humboldt. From Earth it is drawn out into a long, narrow ellipse, but from Orbiter it is seen in its true guise, with a magnificent rill-system on its floor.

In December 1968 Apollo 8 went round the Moon, carrying astronauts Lovell, Borman and Anders. For the first time, human observers had a really close-up view of the lunar scene —and a wild, fascinating scene it proved to be. Since then, as everyone must know, there have been more Apollo flights,

94

eight of which went round the Moon. We have photographs of the surface taken by men as well as by automatic cameras, and Arago's prophecy has come true at last, even though it took a little longer than he had expected when he spoke in the Paris of 1840.

I have not attempted to give a really comprehensive account of the development of lunar mapping; to do so would need many pages. What I have tried to do is to select the main characters in the story. We have the pioneers such as Harriott, Galileo, Hevelius and Riccioli, with their small-aperture telescopes which were so much less effective than modern binoculars; we have the Germans, from Mayer and Schröter to Lohrmann, Beer, Mädler and Schmidt; we have the photographic workers, from Bond and de la Rue to the professionals at Paris and Lick. Then come the last of the mappers who used visual methods together with what photographs they could muster: Elger, Goodacre and Wilkins. Finally, from 1959, individual workers with their own telescopes are, so far as lunar cartography is concerned, superseded by cameras carried in rocket vehicles.

Today the most accurate, detailed and comprehensive maps of the Moon are those drawn up at NASA; the best possible information is used, and there can be little doubt that we have a better knowledge of, say, the floor of the crater Ptolemæus than we have of the thickest jungles of the Amazon. But there are two points to be borne in mind. First, lunar observation itself has not been made obsolete, and there is a great deal to be done yet, as I hope to show in the later chapters of this book. Secondly, and most important of all, the immensely complicated maps of the Moon we use at the present time could not have been drawn without the earlier work of the great selenographers. Schröter, Mädler, Neison and the rest are not mere names from the past; their work lives on.

Chapter Eight

FEATURES OF THE MOON

To the beginner, the first view of the Moon through a telescope gives the impression of tangled confusion. There is so much detail that any attempt to memorize it seems doomed to failure. Yet before very long the impression wears off, and order begins to emerge from the apparent chaos. In particular, the various features sort themselves out into well-defined types, and a few evenings at the telescope – or, for that matter, studying photographs – will be enough for the principal features to be identified.

I am an old-fashioned lunar observer, and the present chapter is given over to old-fashioned observation. Considerations of space-probes will follow later (after all, when I began looking at the Moon the whole concept of lunar travel was equated with perpetual motion, alchemy, and schemes for extracting large quantities of gold from sea-water). For this reason, I propose to show all drawings and photographs with south at the top. I admit that I thought long and hard before deciding upon this, but anyone who wants to be thoroughly 'modern' need do no more than turn the pictures upside-down. The question of what is east and what is west is more difficult. My personal inclination would be to leave things as they used to be, with the well-marked plain of the Mare Crisium toward the western limb and the dark-floored craters Grimaldi and Riccioli to the east; but this was not the decree of the Lunar Commission of the International Astronomical Union – and I feel that the will of the Commission must be followed (particularly, perhaps, as I am myself a member of it). So in this book I will use the astronautical convention: Mare Crisium to the east. The outline map (page 233) should make the situation clear.

Next, there is the question of naming. Some years ago the whole situation became rather out of hand. Riccioli's system of honouring famous men and women was followed, and new names were added—a large number by Beer and Mädler, for instance. The Blagg map of 1935 was an attempt at standardization, but during the 1950s and early 1960s there were several systems in operation, and in particular Wilkins' map included names not found elsewhere. The system I have followed here is in strict conformity with the instructions of the International Astronomical Union, though I have retained obvious English names for mountain ranges—the Apennines and Alps instead of their Latin equivalents, and so on. Anyone using one of the older maps must take care in identification. Thus one very large walled plain on the Moon's limb, beyond the Oceanus Procellarum, is 'Caramuel' on Wilkins' map and various others; the name now officially given to it is 'Einstein', and so Einstein let it be.*

Most observers' maps of the Moon are drawn to mean libration, though on a small scale the differences for any particular limb are minor; and this is the method I have followed here. Actually, libration causes noticeable alterations in aspect. Sometimes the Mare Crisium seems to be very close to the Moon's edge, while at other times it is well on the disk. Areas very badly placed as seen from Earth are difficult to map, and it is only too easy to mistake a foreshortened crater for a mountain ridge. Not until the coming of the Orbiters did we obtain reliable charts of the libration zones—and, of course, the far side of the Moon, which does not concern us in the present chapter.

Newcomers are often nonplussed by the rapid changes in appearance due to the shifting illumination. A peak or crater can alter beyond recognition in only a few hours, simply because we depend so much on shadows. When the Sun is rising or setting over an elevation, the shadow cast is long and prominent, but when the Sun is high above the object the shadow becomes very short, perhaps vanishing completely— just as the shadow of a tree or post will shorten as the Sun rises over it—and as there is no local colour on the Moon, the feature will be almost unidentifiable unless it is definitely darker or brighter than its surroundings. The effect is even more striking with the craters. A walled formation is at its most prominent when it is on or near what we call the terminator.

* The whole matter caused an amazing amount of controversy, some of which, for reasons I have never understood, was acrimonious. Even selenographers are human, with human weaknesses!

The terminator is the boundary between the day and night sides of the Moon. It should not be confused with the limb, which is the Moon's apparent edge as seen from Earth. The limb remains in almost the same position, though it does shift slightly because of the various librations. On the other hand, the terminator sweeps right across the disk twice in each lunation, first when the Moon is waxing ('morning terminator') and then when it is waning ('evening terminator'), so that even an hour's watch will show up definite movement. In Fig. 31, the full, gibbous, half and crescent phases are given, with the limb drawn as a continuous line and the terminator dotted.

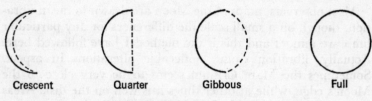

Crescent   Quarter   Gibbous   Full

Fig. 31. The Moon's phases

Owing to the roughness of the Moon's surface, the terminator does not appear as a smooth line. As the Sun rises, the first rays naturally catch the mountain-tops and high areas before the valleys and crater-floors, so that the terminator presents a very jagged and uneven appearance. Peaks glitter like stars out of the blackness while their bases are still shrouded in night, so that the summits appear completely detached from the main body of the Moon; ridges make their first appearance in the guise of luminous threads, while a crater will show its rampart-crests and the top of its central mountain while its floor is still perfectly black. On the other hand, a low-lying area will look like a great dent in the terminator, and take on a false importance for a few hours. Even with a small telescope it is fascinating to watch the slow, steady progress of sunrise over the bleak lunar landscape.

The result of all this is that the features shown on lunar maps cannot be seen properly all at the same time. In fact, it is true to say that full moon is the very worst time for the beginner to start observing, because the limb appears smooth and complete all round the disk, and the shadows are at their minimum. Also there are the bright rays, which drown most of the other details and make the Moon take on the aspect of a speckled, confused circle of light.

Of course the vast dark plains known as seas catch the eye at once, whatever the state of illumination. They take up much of the visible disk, and cover a large part of the western hemisphere—which is why the last-quarter moon, when the western side is shining on its own, is much less brilliant than the first-quarter moon, when the eastern half is visible.

On the Earth-turned side of the Moon (the only one which concerns us in the present chapter) there are nine important seas and a number of lesser ones, though nearly all of these are combined into one great connected system in the same way as our water-oceans, and usually there are no hard and fast boundaries. The names are romantic indeed. Latin is still regarded as the universal scientific language; thus the Sea of Showers becomes 'Mare Imbrium', the Sea of Vapours 'Mare Vaporum' and so on. (*Mare*, I need hardly add, is Latin for sea; plural, *maria*.) We also have a Sea of Nectar (Mare Nectaris), a Bay of Rainbows (Sinus Iridum), an Ocean of Storms (Oceanus Procellarum) and so on—but we have to admit that the names are distinctly inappropriate. Showers, rainbows, nectar and storms are unknown on the Moon.

The seas, with their Latin names and English equivalents, are listed in Appendix VII. The Latin versions are used by astronomers all over the world, and I propose to keep to them here. It is not as though they are in any way difficult; one soon becomes used to them.

We know, of course, that there is no water on the Moon now, and the so-called seas are dry plains without a trace of moisture in them. Analyses of the rocks brought home by the astronauts seem to show that there has never been much water on the Moon; the maria were once lakes of liquid lava, and this has solidified to make the volcanic rock known as basalt. Undoubtedly the maria were viscous well after the adjacent regions had become permanently rigid. There is obvious proof of this in the features bordering them. We can see traces of the old boundary between the Mare Humorum and the Mare Nubium (third quadrant); the mountainous border between the Mare Imbrium and the Mare Serenitatis has been widely breached, while various 'coastal' craters, such as Le Monnier and Doppel-

mayer, have had their seaward walls ruined and levelled. Lava
is a great destroyer.

Many of the major seas are more or less circular, with ramparts that are in places lofty and mountainous. Look, for
instance, at the most impressive of all—the Mare Imbrium
(Fig. 32), in the second quadrant. It seems oval in shape, but
this is because it is foreshortened; really it is almost circular,
hemmed in by four mountain arcs, and large enough to hold
Britain and France put together.

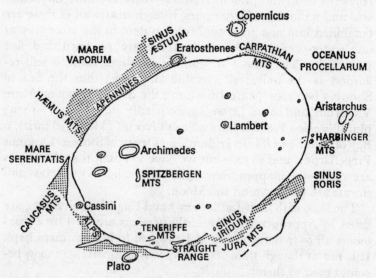

Fig. 32. The mountain walls of the Mare Imbrium

The foreshortening effect is even more striking with the
Mare Crisium, not far from the eastern limb. This is a conspicuous sea, visible with the naked eye and separate from the
main system. It measures 260 miles in one direction and 335
in the other—but the longer diameter is along the east-west
axis, which comes as a surprise to the unwary observer. Orbiter
and Apollo photographs of it taken 'from above' show it in its
true form, and it is distinctive and well defined; but nobody,
apart from the Apollo astronauts, has yet had a direct view of
it as it really is. The third of the almost circular, well-bordered
seas is the Mare Serenitatis, almost exactly equal in area to

Great Britain; it lies in the first quadrant, east of Imbrium. Less perfect, but still basically circular, are the Mare Nectaris and the Mare Humorum, which are smaller structures on the borders of major seas.

There remains the Mare Orientale or Eastern Sea—now, thanks to the IAU resolution, reckoned as being on the western edge of the familiar hemisphere. It is immensely complex, with an essentially circular form and concentric mountain borders. Alone of the really important seas it extends on to the far side of the Moon. The averted hemisphere shows a striking lack of large maria, though there are plenty of craters, valleys and mountains.

The obvious conclusion (obvious to me, at least!) is that there is no basic difference between a large lunar crater and a circular sea, except in scale. Mare Crisium, with its mean diameter of about 300 miles, is not the smallest of the maria. The dark-floored crater Grimaldi measures over 120 miles across, and if it had happened to lie near the middle of the disk instead of near the limb it would probably have been ranked as a minor sea. (On the far side of the Moon there is another example, the dark-floored Tsiolkovskii, though here there is admittedly a most un-sealike central elevation.) The less well-defined dark areas, such as the Oceanus Procellarum—covering two million square miles, much more than our own Mediterranean—are less regular, and are of different type. Probably they are overflows from the circular seas.

J. E. Spurr, an American geologist who was also a pioneer of lunar research, christened the bright upland material 'lunarite' and the dark mare-material 'lunabase'. Lunabase, now known to be basaltic, is not confined entirely to the seas; some craters have their floors covered with it (Billy and Crüger, for instance) and there are also splashes of it here and there in the uplands.

Any casual glance is enough to show that the Moon is a mountainous world. The bright regions, found chiefly in the southern hemisphere, are packed with detail; peaks, craters, valleys and ridges jostle against each other in a wild tangle, and it looks almost as though there is scarcely a square mile of level ground anywhere.

Some of the peaks are lofty, and reach up to well over 20,000 feet above the ground below. But there are quite a number of things to be borne in mind before we can prove that the Moon's mountains are higher than ours, so let us examine the position rather more closely.

Consider the Mare Imbrium, the largest of the circular-type seas. Here there are various mountain borders. The Apennines separate the Mare Imbrium from the smaller, dark-hued Mare Vaporum, and they make a magnificent spectacle when seen at their best, at around the time of half-moon; their peaks cast long, sharp shadows across the plain, and they give the impression of a true range. The highest peak, Mount Huygens, rises to almost 20,000 feet, and there are other crests of at least 15,000 feet; the whole chain is over 600 miles long, stretching from Mount Hadley in the north to the grand crater-ring of Eratosthenes in the south. (The Apollo 15 astronauts landed near Hadley, and their Lunar Rover still stands there, waiting patiently for somebody to go and collect it.)

The Apennines break off near Eratosthenes, and there is a wide gap, so that the Mare Imbrium links up with the even vaster though less perfect Oceanus Procellarum. Then comes another range, the Carpathians, much lower than the Apennines and rising to no more than 7,000 feet anywhere, but over 100 miles long. When the Carpathians come to an end, there is another broad gap in the Mare border.

In the north, the Mare Imbrium is bounded by the Alps, which do not however join up with the Apennines; there is a gap between the two ranges, so that the sea-floor is connected with that of the neighbouring Mare Serenitatis. (There is a difference in level, and also in age.) The Alps are by no means the equal of the Apennines, but they have an interest all their own, largely because of the presence of the remarkable Alpine Valley. Near here, too, is the dark-floored crater Plato, one of the most famous formations on the Moon. Further west lies the Sinus Iridum or Bay of Rainbows, bordered by the Jura Mountains; when on the terminator, so that the Juras catch the sunlight while the Bay is in darkness, the 'jewelled handle' effect is superb. Beyond the Bay, the mountain border is resumed, up to the abrupt ending near the Sinus Roris or Bay of Dews.

The other surviving part of the Mare border is represented by the comparatively modest Harbinger Mountains, which do not make up a definite range, but are better described as clumps of hills.

The general picture is shown in Fig. 32—and one fact stands out at once. The Apennines, Alps, Carpathians, Juras and probably the Harbingers all make up part of the circular wall of the Mare Imbrium. The Mare itself is depressed below the outer country, and it seems that we are dealing with a 'crater' on a tremendous scale. In other words, the mountain ranges are 'crater walls', and not mountains of Earth type.

The same is true of the borders of the Mare Serenitatis, where we have the Caucasus and Hæmus ranges; of the Mare Humorum, where we find the unofficially named Percy Mountains, and so on. Mountain chains on the bright areas are lacking. Even the Altai range, which runs south-westward from the crater Piccolomini, is not properly a chain of mountains, and Spurr's revised name of 'Altai Scarp' is more suitable.

If this interpretation is correct (and let me stress that by no means everyone agrees with me!) then the chief ranges on the Moon are not strictly comparable with terrestrial ranges such as the Himalayas or the Rockies. Yet there are smaller ranges which do not make up the boundaries of seas. Of these, the most intriguing is the Straight Range, near Plato in the Mare Imbrium, a little chain 40 miles long and rising to less than 6,000 feet, but curiously regular, with an abrupt beginning and an equally abrupt end. Also notable are the Riphæan Mountains near the small, bright crater Euclides, which have altitudes of less than 4,000 feet, and may possibly be all that is left of the rampart of a destroyed crater.

Whatever view we take, the Moon's ranges are of the greatest significance in theoretical studies, but we must beware of drawing too close an analogy with the familiar mountain chains of Earth.

There are countless separate peaks on the Moon. They are most numerous in the uplands, but those on the seas are the more impressive. They are not so lofty as the summits of the chain-mountains; all the same, some of them would rank with the well-known peaks of our own world. Look for instance at Pico on the Mare Imbrium, a hundred miles south of Plato— a mountain mass with broad slopes, and foothills studded with pits and craterlets. It may not seem impressive on the map, but the highest crests rises to a full 8,000 feet above the plain, so that it is twice the altitude of Scotland's Ben Nevis. Not far off is another superb mountain, Piton; and there are many other examples.

Less important peaks are very common indeed, and the Moon is dotted with hills, some of them no more than mere mounds. Even the crater-floors are not free of them, and the smoothest parts of the maria are very far from flat, though it is naturally difficult to recognize slight differences in level unless we catch them under a low sun. Well-defined wrinkle-ridges are obvious enough when fairly near the terminator; there is one excellent example crossing the Mare Serenitatis. Many of these ridges seem to be the remnants of the walls of destroyed craters. And, of course, the astronauts have confirmed that the maria are rock-strewn.

On Earth the forces of erosion, such as the wearing-away of heights by wind, water and weathering in general, are constantly at work. But the Moon has no atmosphere, and from the lack of wind and water we might expect the peaks to be much rougher and more jagged. This used to be the popular view, but it has turned out to be wrong. When David Scott and James Irwin, from Apollo 15, drove across the foothills of the Apennines they had a grand view of one of the peaks, Hadley Delta. Scott called it a 'featureless mountain', and the photographs brought home show how right he was.

Galileo was the first to make a serious effort to measure lunar mountain heights. He made use of the obvious fact that the rays of the Sun will catch mountain-tops in preference to the lower-lying areas around, so that a peak may appear as a brilliant point detached from the main body of the Moon. Galileo timed how long the mountain remained lit up when on the night side of the terminator, after which its real distance from the terminator, and hence its height, could be worked out.

This is all very well—in theory. Unfortunately the terminator is so irregular, because of the Moon's uneven surface, that its position cannot be measured properly, and Galileo's results were inaccurate. He believed the Apennines to be about 30,000 feet high; really they are much less than this. A better method is to measure the shadow cast by the peak itself, as

shown in Fig. 33. The position of the peak is known, and so is
the angle at which the Sun's rays strike it at any particular
moment, so that its height relative to the adjacent surface can
be calculated. Of course there are various complications to be
taken into account, but the method itself is straightforward
enough.

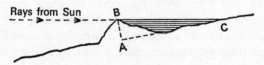

Fig. 33. Measuring the height of a lunar peak (AB)
by the length of its shadow (BC)

Since there are no oceans on the Moon, we cannot relate the
mountain heights to 'mean sea level', but we can at least give
results which are of the right order of magnitude. The one
inescapable fact is that the Moon's mountains are much higher
relatively than those of the Earth, whether we regard them as
mountain ranges or as the ramparts of vast crater-like
structures.

Presumably the seas and the mountain chains must be taken
together when we come to consider how they were formed. At
the moment there are two main theories, diametrically opposed
to each other. Either the regular maria such as Imbrium were
produced by the impacts of giant meteorites, or else—as I
believe—they are of volcanic origin, using the term 'volcanic'
in a very broad sense. No doubt time will tell who is right.

Wherever there are mountains, there will be valleys, and this
is so on the Moon. Some of them are mere passes, while others
are really spectacular. Yet here again we must be wary of
jumping to conclusions. The Rheita Valley, in the Fourth
Quadrant, is long enough to stretch from London to Birmingham, and at first sight looks almost as though it had been
scooped out by a gigantic chisel, but a closer look shows that it
is not a true valley at all. It is a crater-chain from beginning to
end, and there is nothing genuinely valley-like about it.

The Alpine Valley, near Plato, is quite different. Over 80
miles long, it is not a crater-chain; it slices right through the
mountains, and is a superb sight in even a small telescope. It has
no equal on the Moon, and I for one never tire of looking at it.

It is not easy to decide just what is a genuine valley and what is nothing more than a gap between two roughly parallel ridges; the term is a loose one. Many of the valleys, too, resemble that of Rheita in that they are really crater-chains.

Next there are the remarkable features known as domes. As their name suggests, they are surface swellings, and give the impression that they were produced by some internal force which pushed up the Moon's crust without being able to break it. They are not so rare as used to be thought. Robert Barker, an English amateur astronomer, drew attention to some of them in the early 1930s; he was particularly intrigued by a large dome in the crater Darwin, near the limb in the Third Quadrant, and described it as 'bristling with roughness'. Others were soon found, mainly by S. R. B. Cooke in America; using the $12\frac{1}{2}$-inch reflector in my observatory I was able to add some myself. Of course, the space-probe studies have shown a great many more.

Several interesting facts had emerged even before the Orbiter flights. First, the domes are not spread about at random, but occur in clusters. There are several inside the crater Capuanus, not far from the border of the Mare Humorum; other rich areas lie near Arago on the Mare Tranquillitatis, near Prinz in the Harbinger Mountains, and on various parts of the Oceanus Procellarum. One of my favourite domes is the splendid specimen near the little crater Milichius.

Secondly, many of the domes proved to have summit pits, giving them a striking resemblance to true volcanoes. I remember that in 1959 G. P. Kuiper, using the 82-inch reflector at the McDonald Observatory in Texas, announced the discovery of a dozen objects which he called 'extinct volcanoes'. Actually, all these had been included in our earlier published lists, but I would not for one moment quarrel with Kuiper's description of them.

The domes are so gentle of slope that they are not striking even under a low sun, but many can be seen with modest equipment. They also occur on the bright uplands, though the highland domes are naturally even more elusive than those on the grey background of the maria.

Every Earth geologist is interested in faults, and so is every student of the Moon. Examples are not lacking. For instance, the Mare Humorum has been described by V. A. Firsoff as being 'fairly ringed by faults and fractures on all sides', and this is only one of the many heavily-faulted regions.

There are a few particularly outstanding faults; one near Bürg, for instance, in the First Quadrant. But the most striking of all is the remarkable feature known as the Straight Wall, near the edge of the Mare Nubium and not far from the interesting crater Thebit. The name is inappropriate, because the Straight Wall is not straight, and it is certainly not a wall. The surface of the plain to the west is almost a thousand feet lower than on the east, so that the so-called 'wall' is nothing more nor less than a giant fault. It begins at a clump of hills known commonly as the Stag's-Horn Mountains, and ends at a small craterlet 60 miles to the north. Before full moon it shows up as a dark line, since it is casting shadow to the west; it then vanishes, and for some days cannot be identified at all, though the Stag's-Horn peaks can usually be traced. The Wall then reappears as a bright line, with the slanting rays of the Sun shining upon its inclined face.

It used to be thought that the Straight Wall might be almost sheer, so that anyone standing to the west of it would be confronted with a near-vertical cliff. Then, in 1955, some shadow-measurements were carried out by J. Ashbrook in America and by Gilbert Fielder and myself in England; we reduced the angle of slope to about 40 degrees, and the Orbiter pictures have shown that this is correct. The Wall is fairly steep, but certainly not sheer. It is the most perfect structure of its kind, and it is one of the show-places of the Moon. No doubt it will become a tourist attraction in the centuries ahead!

Ridges, too, abound on the Moon, and are of various types. I have already mentioned the so-called wrinkle-ridges (a name due originally, I believe, to Spurr) which are low, snaking elevations of considerable length, seen excellently near the Mare borders. As I pointed out in 1952, the ridges further on the seas are usually the walls of old 'ghost' craters which have been so completely destroyed that only fragments of their borders remain. Yet in general the Moon is criss-crossed with ridges and faults, and this leads on to a phenomenon of tremendous importance—the lunar grid system.

The term 'grid system' was also due to J. E. Spurr. Strictly speaking, Spurr was not an astronomer; he was a geologist (there is a mountain in Alaska named after him) and he did not turn his attention to the Moon until late in his career. He worked from photographs, and so far as I know he seldom looked through a telescope, but he was able to bring all his geological experience to bear upon lunar research.

As Spurr emphasized, the Moon is criss-crossed with faults and ridges. Sometimes the faults are due to almost vertical slippages, while in other cases the movements are more nearly horizontal; generally speaking the ridges form parts of the walls of craters, though often enough the craters themselves are too broken to be properly recognized. There are two principal families of ridges and faults, running in specific directions and almost at right angles to each other. This makes up the main grid system, though it is complicated by the presence of other, less obvious families such as those of the Alps and Caucasus regions.

The very existence of such a grid points to the lunar crust having been subjected to strain over very long periods, and Fielder has also found that small craters in some regions are non-circular, with their longest axes lying in the direction of one or other of the families of the main grid. The general pattern is clear enough—when one knows that it is there. The remarkable thing is that it escaped notice for so long, which may be an example of the age-old cliché about overlooking the obvious.

Finally, before coming to the craters themselves, I must say something about the features known as clefts or rills.* One of the best of them lies near the centre of the disk, near the small but well-marked crater Ariadæus. It can be seen with any small telescope when the lighting is suitable, and it looks remarkably like a crack in the Moon's surface, running for well over 100 miles. Close to it is a curved rill associated with the 4-mile crater Hyginus. Most spectacular of all is the lovely winding valley-cleft running out of Herodotus, the companion-crater of the glittering Aristarchus near the junction of the Mare Imbrium and the Oceanus Procellarum. It starts inside Herodotus, broadens out into a formation which has been nick-named the Cobra-Head, and winds its way across the plain. Then, too, there are whole systems of rills, such as that in the

* The German spelling, *rille*, is often used.

region of Triesnecker—again in the general area of Hyginus and Ariadæus; and some craters, such as Gassendi on the edge of the Mare Humorum and Alphonsus in the central chain of which Ptolemæus is the senior member, have elaborate cleft-systems on their floors.

In pre-Space Age days there were some ideas about rills which we now know to be wrong. R. B. Baldwin, in 1949, described the Triesnecker clefts as 'irregular cracks, very jagged in appearance, which seem to be bottomless'. In fact they are not more than a mile or two deep at most, whereas they are up to three miles wide. They are not jagged, and they are not true cracks; there is no doubt that they are collapse features, and there is no analogy with our river-beds. No water has ever flowed in them. Their distribution is not random; some parts of the Moon are riddled with rills, while other areas lack them completely. Their origin must be bound up with that of the craters—and, of course, with that of the grid system.

Some of the so-called rills are crater-chains, at least in part. The Hyginus Rill, for instance, has crater-like enlargements along it, and many years ago D. W. G. Arthur, one of the few British amateurs who turned professional and went to America to work in a lunar team, published a drawing in which he made the craters extend from one end of the feature to the other. This is not quite correct, as the Orbiters have shown, but the tendency is there, and there are other examples of the same kind of structure. Remember, too, that one rill has been studied from close range. From Apollo 15, David Scott and James Irwin drove almost to the edge of the vast rill near Hadley in the Apennines.

So much, then, for the seas, the mountains, the domes and the peaks, the faults, ridges and rills, and the grid system. Up to now I have only touched upon the craters of the Moon—and yet these craters are the dominating features of the lunar landscape. Without them, the Moon would look strange indeed.

Chapter Nine

THE CRATERS OF THE MOON

WHO HAS NOT HEARD of the craters of the Moon? They dominate the whole lunar scene; no area is free from them. They are found on the plains, in the rugged uplands, on mountain-tops and on the floors and walls of larger structures. They range in size from vast enclosures well over 100 miles in diameter down to tiny pits, so small that from Earth they cannot be seen at all. As the astronauts walked (or drove) across the Moon they found crater-pits everywhere.

Strictly speaking, the term 'crater' is rather misleading, since it conjures up the idea of a deep, steep-sided hole, and the principal structures are not in the least like this. Also, one thinks instinctively of craters such as that on the summit of our own Vesuvius; and although there are plenty of small features on the Moon which, in the words of the late G. P. Kuiper, look like extinct volcanoes, the large enclosures are utterly different. Various schemes of classification have been proposed. For instance, the large formations with relatively low ramparts and without central peaks have been called 'walled plains', smaller features 'crater-rings' and so on. As this is a purely observational chapter, I propose to take the easy way out and call all the walled structures 'craters', but I must stress that they show a very wide range in both scale and form.

The scale is perhaps the most striking thing about them. Consider Theophilus, one of the most impressive craters on the Moon, with high walls and a massive central mountain complex. Its diameter is 64 miles, and it has well-marked borders. Bailly, much less deep, is around 180 miles in diameter. Transfer either of these to England, and there would not be much room to spare, as shown here (Fig. 34). There is nothing on Earth to match craters of this kind; our largest structures would make a very poor showing on the Moon. The famous Meteor Crater in Arizona and the volcanic Hverfjall in Iceland, for instance, are less than one mile across, so that if they lay on the Moon they would certainly not be honoured by

being given separate names.
Then, too, appearances can
be deceptive. When a crater
is seen on the terminator, so
that its floor is filled with
shadow, it seems immensely
deep, like the proverbial 'bottomless pit'. Look for instance
at the picture of Ptolemæus,
in Plate XIII. It is hard to credit
that the walls are nowhere
more than 4,000 feet high,
whereas the diameter of
Ptolemæus is almost 100 miles.
A large lunar crater is not a
deep hole; it is more like a
shallow saucer (Fig. 35). As

Fig. 34. The area of England and Wales which would be occupied by the crater Bailly

yet no astronauts have been to Ptolemæus, but an observer
standing in the middle of the formation would be unable to see
the walls at all, for the excellent reason that they would be well
below his horizon.

In some cases the depths are greater. One of the deepest on
the Earth-turned hemisphere is Newton, near the south pole,
where the walls rise to well over 20,000 feet above the lowest
part of the interior, and there are parts of the floor which never
receive any sunlight at all. Yet even Newton is comparatively
shallow when we compare its depth with its diameter.

The smaller craters are somewhat different in form. As a
typical example, consider Theætetus, on the Mare Imbrium
(second quadrant). It is 32 miles across, so that its area is
greater than that of the Isle of Wight, and the walls rise to
7,000 feet above the floor. This sounds quite impressive; but the
floor itself is depressed 5,000 feet below the plain, so that anyone
standing on the Mare Imbrium and looking at the outer rim
would be confronted only with modest elevations much lower
than Scafell. Neither are the slopes steep, though it is true that
astronauts have found walking uphill very exhausting—of
course, space-suits are not the most comfortable of garments for
long treks! Even with really small craterlets the slopes are not
precipitous, and the idea of a lunar crater as a gaping hole in

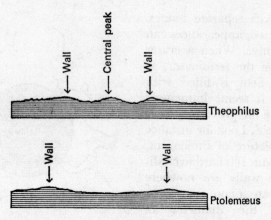

Fig. 35. Crater profiles: Theophilus and Ptolemæus

the surface, banked by mountains rising almost sheer from the shadowed depths, is very wide of the mark.

There is one common factor. All craters, large or small, are basically circular, even though they may have been battered and distorted by later activity; in this they resemble the circular seas, such as Mare Imbrium and Mare Crisium. As seen from Earth, the craters away from the centre of the disk appear as ovals, due to foreshortening, but the Orbiter and Apollo pictures give the true story. Look for instance at two photographs of the regular, dark-floored Plato, which is 60 miles in diameter and is instantly recognizable whenever it is in sunshine. The Earth-based picture shows it elliptical; the Orbiter view, taken from almost overhead, proves it to be virtually a perfect circle.

The gradation from circle to long, narrow ellipse is very evident. Consider Pythagoras in the second quadrant, Clavius in the rough southern uplands, and Arzachel in the Ptolemæus chain. All are circular—but only Arzachel looks it. Clavius is 146 miles across, so that its area is greater than that of Switzerland; a string of smaller craters crosses its floor, and the walls are some three miles above the depressed interior. Clavius seems decidedly oval, because it is closer to the limb than the centre of the disk, while Pythagoras is so drawn-out that its interior details are hard to see at all. Were it better placed, it would be a most imposing object.

The largest craters—those known conventionally as walled plains—must also be the oldest. Many of them are now so broken and ruined that they are scarcely recognizable. Janssen in the fourth quadrant, not far from the Mare Australe or Southern Sea, must once have been a noble formation, with high continuous walls towering to thousands of feet above its sunken floor, but it has been so roughly treated by later disturbances that it is now no more than an enormous field of ruins, broken by craters, ridges, pits and rills, with its walls breached in dozens of places and completely levelled in some. Only when it is right on the terminator, and filled with shadow, does it give a slight impression of its former self.

The term 'field of ruins' has also been applied to Bailly, on the south-western limb of the Moon. It is actually the largest-named crater on the Earth-turned hemisphere, but it is so badly placed and foreshortened that only the space-probe pictures can show it well. Were it darker-floored and more central on the disk, it would certainly have been classed as a minor sea rather than as an outsize walled plain.

However, quite a number of large structures have managed to escape relatively unharmed. Clavius is a case in point. Admittedly it contains several craters which are large in their own right, but the walls are fairly regular, and when the Sun is rising or setting over it the result is an apparent dent in the terminator which can be seen without any optical aid at all.

Many of the walled plains, particularly the damaged ones, are hard to identify when the Sun is high over them. The exceptions are those with dark floors, such as Plato, which Hevelius called 'the Greater Black Lake'. Grimaldi and Riccioli, near the western limb, are also extremely dark, and there are some smaller formations with similar lunabase floors; Billy and Crüger, both in the third quadrant, are examples.

There is a tendency for the walled plains to be arranged in lines and groups, a point to which I will return later; there is nothing random about their distribution, though we now know that on the Moon's far side the situation is less clear-cut. One of the most impressive of the chains lies very close to the apparent centre of the disk. Here we have Ptolemæus, Alphonsus and Arzachel, which are magnificent when near the terminator though decidedly obscure near full moon.

Ptolemæus has a reasonably darkish floor, with few conspicuous craterlets and no central mountain. Alphonsus, 80 miles across, is the middle member of the chain; there is a reduced central peak, and on the floor there is a fine system of rills, photographed in detail not only by the Orbiters and Apollos but also by the last of the Ranger probes, No. 9, which crash-landed not far from the central peak. The southern member of the trio, Arzachel, is smaller and deeper than Alphonsus, with a considerable central elevation. We have here an excellent gradation between walled plain and peaked crater, adding force to my contention that the structures are of essentially the same type and origin.

Coming now to the generally smaller formations, we find that central peaks are common, though by no means the universal rule. Some craters have floor-mountains which rise to thousands of feet, though they never equal the height of the surrounding rampart; others may have lower, many-peaked central elevations, and sometimes the so-called central mountain is little more than a mound. Then, too, there are formations which have smaller craters centrally placed on their floors, and there are many examples of true concentric crater-rings. Taruntius, near the Mare Crisium, and Vitello on the edge of the Mare Humorum are good examples.

One of the finest of all the regular craters is Theophilus, in the fourth quadrant near the edge of the Mare Nectaris, 64 miles in diameter and over 14,000 feet deep, with a magnificent central elevation and massive walls. It has broken into a second formation, Cyrillus, which is about the same size, but whose walls and central peak are much lower; to the south lies the third member of the chain, Catharina, where the floor is very rough and lacks a central mountain. On the Mare Imbrium we have the superb trio made up of Archimedes, Aristillus and Autolycus, where the same sort of gradation in type is shown; and there are others almost equally worthy of note. But among all the Moon's craters perhaps the most impressive is Copernicus, on the Oceanus Procellarum. It has been nicknamed 'the Monarch of the Moon', and with good reason.

Copernicus has massive walls rising in places to 17,000 feet above the inner amphitheatre. The distance right across the crater, from crest to crest, is 56 miles, but the true 'floor' is only

40 miles across, since the rest is blocked with rubble and débris produced by huge landslides from the ramparts. The central heights are made up of three distinct, multi-peaked masses, while lower hills and mounds litter the floor. There is nothing smooth about the inside of Copernicus. The walls are terraced, a not uncommon feature in lunar craters but particularly well-marked here. The outer slopes of the wall, which are comparatively gentle, are lined with ridges and valleys which radiate outward. Spurr believed that these features were due to water pouring down the outer slopes from the central orifice, much in the way that narrow channels are formed in a sand-bank when water is running down it. I think we may safely reject this explanation now, because the rock analyses have shown that there has never been any appreciable water on the Moon; but a flow of some kind or other may have been involved, and it is true that other craters, such as Aristillus, show similar gullies.

In 1966 Orbiter 2 photographed Copernicus from a mere 28 miles above the Moon's surface. The crater, approximately 150 miles away, was shown obliquely, and the result was termed the 'Picture of the Century'; it was the most magnificent lunar view obtained up to that time, though of course it has been surpassed since. Another Orbiter picture showed flows in the inner wall—due probably (at least in my opinion) to volcanic lava. And one peak in the group of central elevations looks so strikingly like a true volcano that few people believe it to be anything else.

These fine details are not visible from Earth, even with our best telescopes, but a modest instrument is enough to show that Copernicus is a sight never to be forgotten. Moreover, it is the focal point of a system of bright rays second in importance only to those of Tycho—though admittedly they are of rather different type.

Copernicus, with a probable age of perhaps a thousand million years, is one of the youngest of all the major craters, and has escaped damage, but other formations have not been so lucky. Those on the 'coasts' of the maria have had their sea-ward walls battered down and levelled, so that they have been turned into huge bays; Fracastorius on the border of the Mare Nectaris and Le Monnier on the edge of the Mare Serenitatis are good examples, and the same may be true of the famous Sinus Iridum or Bay of Rainbows. In some cases the ruins of a seaward wall can still be seen, and even, as with Hippalus on the border of the Mare Humorum, the wreck of a central mountain.

The Sinus Iridum, most splendid of all the bays, leads off the Mare Imbrium. The ground level drops gradually to the west, and low, discontinuous remnants of the old east wall can still be traced between the two jutting capes to either side of the strait which separates the Bay from the main Mare. When the terminator passes close by, the mountain peaks of the western border (the Juras) catch the light, producing the unique 'jewelled handle' effect.

Old craters right on the seas have been even more reduced, and have been drowned by the Mare material, so that they now show up as ghosts—marked sometimes by low, discontinuous walls, sometimes by ridges, sometimes by nothing more than a slight change of tint on the plain.

Look, for instance, at Stadius, not far from Copernicus. It is large enough to hold the whole county of Sussex, and must once have been a noble formation, but nowadays it is in a sad state. The Mare material has flowed across it, breaching the ramparts and leaving them shattered and ruined. The loftiest summits are no more than a couple of hundred feet high, and for long stretches the wall cannot be traced at all, while the amphitheatre is speckled with numerous tiny pits. Unless caught under very oblique lighting, it is difficult to find. Its neighbour Eratosthenes, only about a hundred miles off, has escaped completely, and must have been formed much later.

The arrangement of the lunar craters is interesting. Like the great walled plains, they tend to line up, and there are also pairs and groups. With some twins the two craters are quite separate, as with Aliacensis and Werner, which lie in the fourth quadrant close to the vast chain of which Walter is the senior member; in other cases the second formation has broken into the first, as with Steinheil-Watt in the fourth quadrant and the Sirsalis pair in the third. When this happens, the intruding formation is almost always slightly the smaller of the two. In fact, the rule of smaller craters breaking into larger holds good in over 99·9 per cent. of all known cases.

It is tempting to go on describing crater after crater, since each has its own special points of interest. I shall have more to say about some of them later—notably Aristarchus, which is a mere 23 miles in diameter, but is the most brilliant feature on the entire surface, so that it shows up even when illuminated only by earthshine. There is plenty of variety on the Moon.

Craterlets, with diameters ranging from more than a dozen miles down to a few yards or even less, pepper the whole Moon. Some are true miniatures of larger craters, even to the central hills; others are pits, with depressed floors but virtually no walls rising above the outer level (Spurr called them 'blowhole craters'). Probe pictures, and above all the photographs taken by the astronauts on the surface itself, show that there are countless craterlets much too small to be seen from Earth.

The lining-up is even more marked with the small formations than with the larger ones. Many of the so-called rills are made up basically of small craters which have run together, often with the loss of their dividing walls. Others, such as the Hyginus Cleft, are crater-chains in part, though it is true that there are plenty of rills which show no trace of crater-like enlargements. Crater-chains are very common indeed, and there is nothing surprising in this. We have a complete series from the 'giants' down to the 'strings of beads'.

Here and there we come across real lunar freaks. Perhaps the most remarkable of all is Wargentin, close to the large walled plain Schickard in the third quadrant. Its floor is not sunken, but is raised above the outer surface by about a thousand feet. What may have happened is that some blockage caused the molten magma to be trapped inside the amphitheatre when the crater had just been formed, so that instead of subsiding and draining away, as usually happened, the magma solidified where it was. If so, the true floor of Wargentin is hidden, and all we can see is the top of the lava-lake. In places the floor is level with the top of the rampart, but in other areas there are still traces of a wall—one segment rises to as much as 500 feet. Still, the general impression is that of a flat-topped plateau, not a conventional crater. Wargentin is large enough to hold the whole of Lancashire, and it is a pity that it is not more central on the disk, as there are no other plateaux anything like as

117

large. Various smaller specimens exist, but they are neither so perfect nor so prominent.

Finally there are the bright rays, which dominate the whole scene when the Moon is near full. Unlike most other details, they are best seen under a high light; they are very obscure when near the terminator, and become conspicuous only when the Sun has risen to a considerable altitude over them. Of the many ray-systems on the surface, two stand out as being incomparably more splendid than the rest: those of Tycho and Copernicus.

Tycho is a well-formed crater in the southern uplands, 54 miles across, with high terraced walls and a central mountain complex. Magnificent though it is, Tycho lies in so crowded an area that it would not be outstanding were it not for the rays. When it first emerges from the long lunar night it seems to be a perfectly normal bright crater, but gradually the rays come into view, until by full moon they dominate the whole of the southern part of the disk. There are dozens of them, streaking out in all directions from Tycho as a focal point; they cross craters, plains, peaks and valleys, uplands and maria, rills and pits without showing obvious deviation.

Non-astronomers often think that Tycho marks the south pole of the Moon. This is wrong; in fact Tycho lies some way from the polar point.

Strange to say, the rays cannot be followed right into Tycho. There is a ray-free area round the rampart, showing darkish under a high light, where the streaks stop short; neither do they radiate from the exact centre of Tycho, since many of them are tangential to the walls. Yet there can be no doubt that the crater and the rays are closely associated. One ray stretches right down beyond the Mare Serenitatis, passing close to the bright little crater Bessel, and there has been a great deal of discussion as to whether this is one long, genuine 'Tycho ray' or whether it is renewed along its course.

Needless to say, Tycho was photographed from the Orbiters, and the great roughness of the floor was confirmed. Then, in January 1968, the soft-landing Surveyor 7 came down on the outer slopes of the crater, and sent back plenty of pictures; chemical analysis of the region was also carried out, and it was found that there was an unexpected abundance of aluminium, together with rather less iron than in some of the other sampled sites. Surveyor, its power long since exhausted, is still standing where it touched down. Some day, no doubt, an expedition will visit Tycho, collect Surveyor 7, and remove it to some lunar museum.

The rays associated with Copernicus are different from those of Tycho. They are not so luminous, and at full moon, when they are at their best, they appear less brilliant than the gleaming crater-ring of Copernicus itself. Neither are they so long or so regular as those of the Tycho system, though they spread widely over the surrounding plain.

Here and there over the disk other ray-centres can be made out: Kepler on the Oceanus Procellarum, Olbers close to Grimaldi in the far west, Anaxagoras in the north, and so on. Some craters have rays which are so dark as to be scarcely detectable. We also find small craters surrounded by bright patches—Euclides, near the Riphæan Mountains, is an example—and craterlets with short ray-systems. Near full, the various rays confuse the whole lunar scene so thoroughly that even the practised observer may have trouble in finding his way about.

The rays are not continuous white streaks, but show definite structure. They are surface deposits, not cracks nor elevations. I would hesitate to say that we really understand them, even though Apollo astronauts have visited areas affected by ray material.

This, then, is the lunar picture; a scene of apparent chaos which, upon close examination, turns out to be not quite so chaotic after all, and where craters and peaks mingle with valleys, faults, clefts and mountain chains. The fact that twelve men have now walked upon the surface does not make the Moon one whit the less interesting—or the less puzzling.

Chapter Ten

ATMOSPHERE AND LIFE?

THIS CHAPTER MUST, I think, be essentially historical. When I wrote the first edition of *Guide to the Moon*, a quarter of a century ago, lunar travel still lay in the indeterminate future, and it was thought quite possible that there was an atmosphere around the Moon—excessively tenuous, of course, but not completely negligible. The existence of primitive life-forms was generally discounted, and the fascinating 'Selenites' of fiction had been reluctantly jettisoned by all scientists, but there was still no positive proof that the Moon must be utterly sterile.

Nowadays the situation is different. The lunar atmosphere really is negligible, and there is no life on the Moon of any kind. Yet the story of how these conclusions were reached is an interesting one, well worth re-telling.

First, then, let us consider the lunar atmosphere—or lack of it. Even in the earlier part of the nineteenth century astronomers such as Schröter and Herschel still believed the density to be appreciable, but there is an easy way to show that any atmosphere round the Moon must be far less dense than ours. This method involves what are known as lunar occultations.

As the Moon travels across the sky it often passes in front of stars, hiding or occulting them. There are a few really bright stars which lie in the Zodiacal band, and can therefore be occulted; Antares in the Scorpion is one, Aldebaran in the Bull another. Occultations of fainter stars are common enough, and amateurs carry out valuable work in observing them, using nothing more than a telescope plus a reliable stop-watch. What is needed is the exact time when the star is hidden by the onward-moving limb of the Moon. This gives the precise position of the Moon at that moment—because the positions of the stars in the sky are known much more accurately than that of the ever-shifting Moon.

Timings can be very definite, because an occultation is virtually instantaneous. Before occultation, the star is seen shining steadily; its disappearance behind the Moon's limb is as sudden as the flicking-out of a candle-flame in a gale. One moment the star is there, the next it is not. Reappearance at the opposite limb of the Moon is equally sudden.

Now consider what would happen if the Moon were surrounded by a dense atmosphere. For some moments before being hidden the star's light would be coming to us after having passed through the lunar atmosphere, and the star would flicker and fade. Venus has a very extensive atmosphere (much denser than ours) and this makes a star fade very obviously before occultation. I had a good view of this in 1959, when, for the first time in many years, Venus occulted a brilliant star, Regulus in Leo. Together with Henry Brinton, and using the 12½-inch reflector at his observatory in Selsey, I recorded a perceptible fading of Regulus which was undoubtedly due to the dimming of the starlight by Venus' atmosphere. All other observers who watched the occultation under good conditions saw the same effect, and the space-probe results of the 1960s and 1970s have since told us that our resulting estimate of the depth of Venus' atmospheric mantle was very near the truth. The same sort of effect was seen when Mars occulted the third-magnitude star Epsilon Geminorum on April 8, 1976—though on that occasion I was unable to obtain any accurate results because from my observatory Mars was only just above the horizon.

An occultation of a star by the Moon is interesting, particularly when the Moon is waxing—since the occultation then takes place at the dark limb, which cannot be seen unless lit up by earthshine. When the Moon passes through a star-cluster such as the Pleiades, half a dozen naked-eye stars may be occulted over a period of a few hours. A relatively small telescope is quite useful for adequate timings to be made.

There is another method of attack, also. If a belt of atmosphere lay round the lunar limb, it would be expected to bend or refract the light-rays coming from the star just before immersion (you can demonstrate refraction simply by shining a torch through a tank of water; the beam is obviously bent). The effect would be to keep the star in view for a little longer than would otherwise be the case, so that the occultation would take place later than predicted. Reappearance at the opposite limb would be slightly early, so that the whole occultation would last for a shorter time than as forecast by theory. The amount of the time-difference should give a reliable key to the density of the lunar atmosphere responsible for it.*

This sounded very plausible, and it only remained to measure the time-discrepancies. For reasons which we now know, this proved to be impossible. Sir George Airy, who was Astronomer Royal between 1835 and 1881, believed that the effect was great enough to be detected, but later astronomers, using more sensitive equipment, did not agree. This was hardly surprising; the lunar atmosphere is so negligible that the theoretical times for immersion and emersion of a star are correct.

Now and then, however, odd effects are seen, and I observed one myself on 19 March, 1972, when I was timing occultations in the famous star-cluster of the Pleiades or Seven Sisters, using my 12½-inch reflector. Instead of vanishing with the usual abruptness, one star quite definitely faded out. I was taken aback, and would probably have dismissed the observation as being faulty had it not been confirmed by a colleague, L. Anslow, who was within a few yards of me and was using my portable 4-inch refractor. We later established that the star was a binary—that is to say a system made up of two stars, so close together as to appear as one object. This explained the fading, since the two components were hidden at fractionally different times.

Research into the literature showed that other fading occultations had been seen now and then, and it does not seem that all the stars concerned are binaries; but whatever the cause may be, it is certainly not due to a lunar atmosphere, or even a local emission of gas from below the Moon's surface. The density of the gas required would be impossibly high.

When a planet is occulted by the Moon the situation is naturally quite different, because a planet shows a disk, and both disappearance and reappearance are gradual. W. H. Pickering, who was an energetic observer of the Moon even if

---

* Refraction makes it possible for us to see the Sun and Moon before they really rise, because refraction 'lifts them up' from below the horizon. On some occasions the Sun and the full moon may be seen simultaneously, just above opposite horizons. Moreover, both the Sun and the Moon look flattened when very low down, because the bottom part of the disk is the more strongly refracted.

he seems to have been less successful as a theorist, watched an occultation of Jupiter in 1892, and recorded a dark band crossing the planet's disk, tilted at an angle to the famous Jovian belts. Pickering believed this to be due to absorption in a lunar atmosphere, and he confirmed it on other occasions. He claimed that the dark band was seen only when the planet passed behind the Moon's bright limb; when occultation took place at the dark limb no band was seen, and Pickering concluded that the atmosphere responsible for it was frozen during the lunar night. Others who recorded dark bands at planetary occultations were two of Pickering's colleagues, Barnard and Douglass, both of whom—Barnard particularly—were known to be magnificent observers.

From this kind of result, Pickering worked out a density of the lunar atmosphere as 1/1800 of that of the Earth's air at sea-level. This value was not only too great, but impossibly too great. Neither has the phenomenon been seen with any consistency, and it has never been photographed, which makes one suspicious; the human eye is very easily deceived. Nevertheless, there are some more recent observations which remain to be explained. When Saturn was occulted, on 2 March, 1974, one of the British Astronomical Association observers, L. E. Fitton, used his $8\frac{1}{2}$-inch reflector to report that when the lunar limb touched the bright inner ring of the planet 'a very fine red line appeared at the dark outline of the Moon's limb where it crossed the ring system . . . the red line persisted until about three-quarters of the globe had been occulted, when it quite suddenly vanished and did not reappear at any time'.

A report of that kind, coming from an observer of known skill and experience, cannot be discounted, but whether it can be attributed to any lunar atmosphere is another matter. In my view, at least, it cannot. Once again the density of any gas capable of producing such an effect would have to be much greater than we can accept.

Another line of investigation was tried out when the Moon occulted the Crab Nebula in Taurus, which is known to be the wreck of a star which exploded in a supernova explosion long ago, and which is a source of radio waves. (The supernova outburst was seen by Chinese and Japanese astronomers in the year 1054, but since the distance of the Crab is 6,000 light-years the actual explosion took place in prehistoric times.) Radio astronomers at Cambridge studied occultations of the Crab to look for effects roughly analogous to those for visual observations, and initially concluded that there were signs of a slight atmosphere round the Moon, but these results too were unconfirmed.

Neither are there any marked twilight effects on the Moon. Reports have been made from time to time, but proof is signally lacking. I may well be prejudiced here, since I have looked for twilight effects at the cusps of the crescent Moon more times that I can count, with completely negative results.

Two Russian astronomers, V. Fesenkov and Y. N. Lipski, attacked the problem in the 1940s. If there is any light coming from diffusion in a thin lunar atmosphere, it should have special qualities, and therefore be detectable with sensitive equipment. Fesenkov, in 1943, had no luck at all, and concluded that the Moon's atmosphere could not have a density greater than one-millionth of our own. Six years later Lipski made a new investigation, and obtained different results, announcing a value of 1/20,000. This was confirmed in 1952, and in 1953 Lipski published another paper in which he raised the density to 1/12,000—agreeing well with earlier work carried out by Bernard Lyot in France.

Yet it was not long before doubts crept in. Lyot and his colleague Audouin Dollfus, working at the Pic du Midi Observatory in the Pyrenees, failed to confirm Lipski's results. After Lyot's sudden and tragic death Dollfus carried on the work, and decided that the lunar atmosphere was too tenuous to be detected at all. This would mean a density not greater than one thousand-millionth of that of the Earth's air at sea-level.

What, then, about the possibility of seeing meteors in the Moon's atmosphere? With a ground density as high as the value given by Lipski, it could be possible. In 1952 I discussed the possibility with Dr. E. J. Öpik of Armagh Observatory, a world authority on meteoritic phenomena. He then wrote as follows:

Lunar meteors are quite probable. Considering the surface gravity of the Moon, which leads to a six times slower decrease of atmospheric density with height, the length of path and duration of meteor trails on the Moon will be six times that on the Earth, if a lunar atmosphere about 1/20,000 to 1/100,000 of the density of the terrestrial atmosphere exists. At the same time, meteors of the size of fireballs will penetrate the lunar atmosphere and hit the ground. The average duration of a meteor trail on the Moon would be two to three seconds (as against half a second on Earth), because all meteors which can be observed on the Moon from such a distance must be large fireballs. The average length of the trail would be 75 miles, about 1/30 of the Moon's diameter, so that meteors would be very short and slow objects.

Öpik added that with a 12-inch telescope it should be possible to record an average of one lunar meteor for every eight hours' work.

All this seemed very reasonable at the time, and careful searches were made. My own were completely negative, but in the United States many streaks were recorded by W. H. Haas and his colleagues in the Association of Lunar and Planetary Observers. The average trail-length worked out at about 75 miles, agreeing excellently with Öpik's estimate. Haas even calculated the probable diameter of an object whose trail he recorded in 1941; assuming it to have been a genuine lunar meteor it would have been 600 feet across.

Now that we know, from the Apollo missions, that the Moon has no atmosphere capable of producing meteors, we must take a very critical look at these results. Whatever the observers saw, they can have recorded no meteors over the lunar surface, and the whole episode may well be a warning that it is very easy to 'see' what one half-expects to see. There are only two possible explanations. Either the American watchers were deceived by tricks of the eye, or else they were seeing phenomena of a different sort—which, in view of the descriptions given, does not seem very likely.

I have given the story of the lunar atmosphere in some detail because it is, I think, interesting to look back and see just how our ideas have changed over the years.

The last word was said in December 1972, when the astronauts of Apollo 17—Cernan and Schmitt—set up an instrument known as a mass spectrometer on the lunar surface. An excessively tenuous atmosphere was found, but it is not of the same kind as ours, and is what is termed a 'collisionless gas', in which the various atomic constituents move around more or less freely. Gases detected included hydrogen, helium, neon and argon. The argon was found to be more in evidence at lunar dawn, indicating that it was frozen during the bitterly cold night and was released into the atmosphere at the sunrise terminator. Probably the atmosphere is derived from the solar wind—streams of low-energy particles sent out by the Sun constantly and in all directions, so that some of the particles must reach the Moon. But the overall density of the lunar atmosphere is so low that for all practical purposes we may safely say that the Moon really is an airless world, and at least one long-standing problem has been cleared up at last.

There is no mystery about the lack of atmosphere. The Earth's escape velocity is 7 miles per second, and this is why our air is relatively dense; the atoms and molecules simply cannot move quickly enough to break free. With the Moon, the escape velocity is a mere $1\frac{1}{2}$ miles per second, which leads to a very different state of affairs. Even if there had once been a dense atmosphere the Moon's pull could not have retained it, and the atmosphere would have leaked away into space comparatively quickly. This is not to suggest that an Earth-type atmosphere has ever existed there; I suspect that it has not. There is, of course, much to be learned from conditions on other worlds. Jupiter (escape velocity 37 miles per second) has been able to hold on to all its gases, even the lightest of all, hydrogen; Mars ($3\cdot1$ miles per second) has only a thin atmosphere, made up chiefly of the heavy gas carbon dioxide; Mercury ($2\cdot5$ miles per second) has almost no atmosphere at all.

Without atmosphere, can we expect life? The instinctive answer is 'No', and we may now be sure that this is correct, since the lunar samples brought home for analysis show a total lack of any organic material either past or present. So once again I propose to delve back into history, and any reader who is anxious to come on to modern investigations has my full permission to pass on immediately to Chapter 11!

We should begin, I feel, by narrowing down what is meant by 'life'. Our knowledge of the basic nature of life is still very fragmentary, and it has often been suggested that there may well be creatures of entirely alien pattern—made up of gold, perhaps, and breathing pure hydrogen. Beings of this intriguing type are well known to story-tellers, who generally term them

BEMs (Bug-Eyed Monsters) and scatter them indiscriminately on any worlds which have not actually been explored by space-craft. Let us admit at once that we cannot rule out the possibility of BEMs. Shakespeare's lines 'There are more things in Heaven and earth, Horatio . . .' hold good in science, as in everything else. A colleague of mine once said that he could not deny the possibility of an intelligent extra-terrestrial being who looked like a cabbage and squeaked like a mouse. He did not think it likely, but it was not absolutely out of the question—and of course he was right. Yet when we start to consider totally alien forms, speculation becomes both endless and pointless. We know a good deal about the way in which living matter is built up, and all available evidence indicates that BEMs do not exist anywhere. If they do, then all of our modern science is wrong; and this I am not prepared to believe when there is not a shred of evidence in favour of it. So we can justifiably discount alien forms, and confine ourselves to discussing life *as we know it*.

Various conditions are necessary for our sort of life. There must be a reasonably even temperature, an atmosphere containing oxygen, and a supply of moisture. On the Moon, the lack of atmosphere means that the temperature-range is extreme; at noon on the lunar equator a thermometer would register over 200 degrees Fahrenheit, while the midnight value is of the order of minus 250 degrees. Without atmosphere there can be no water, and it is now certain that the lunar 'seas' were never water-filled. In every way, then, the Moon is hopelessly hostile. This seems so obvious today that we tend to forget that less than two centuries ago leading astronomers were quite ready to believe the Moon to be a world capable of supporting intelligent beings.

The idea of a populated Moon is very old, and once it was realized that the Moon is a rocky globe it was tacitly assumed to be inhabited. Even the invention of the telescope did not bring about a prompt change of view. In 1634 came the posthumous publication of the famous *Somnium* or 'Dream', written by no less a person than Johannes Kepler (the man who first proved that the Earth moves round the Sun in an elliptical path), in which various weird and wonderful life-forms were described. Quite probably Kepler was deliberately letting his imagination run riot, but he and others of the period were very ready to believe in a Moon with extensive oceans and a dense atmosphere.

By 1800 the idea of oceans had been abandoned, but it was still thought that there might be air and water, and that life survived on the surface. Schröter certainly believed so, and he was supported by the most famous astronomer of the day, William Herschel.

Herschel, the Hanoverian musician who became official astronomer to King George III, is best remembered for his discovery of the planet Uranus, in 1781, but his most important contributions were in the field of stellar astronomy (he was one of the first men to give a reasonable picture of the shape of our star-system or Galaxy). As an observer it is possible that he has never been equalled, and between 1781 and his death, in 1822, every honour that the scientific world could bestow came his way. His views about life in the Solar System were, then, rather surprising. He thought it possible that there was a region below the Sun's fiery surface where men might live, and he regarded the existence of life on the Moon as 'an absolute certainty'. In 1780 he wrote to the then Astronomer Royal, Nevil Maskelyne, as follows:

Perhaps conclusions from the analogy of things may be exceedingly different from truth; but seeing that our Earth is inhabited, and comparing the Moon with this planet; finding that in such a satellite there is a provision of light and heat; also in all appearance, a soil proper for habitation fully as good as ours, if not perhaps better—who can say that it is not extremely probable, nay beyond doubt, that there must be inhabitants on the Moon of some kind or other?

Maskelyne was not in the least convinced, and it is on record that later, when Herschel said much the same thing in a paper about lunar mountains, the Astronomer Royal deleted the offending paragraph before passing the paper for publication. Yet the idea of moon-men was certainly not dismissed out of hand.

Schröter's views were not so extreme, but he too was reasonably sure that the Moon must be populated. He knew that the lunar atmosphere is thin, but he grossly over-estimated its density, and even considered that some of the features visible on the Moon might be artificial. This last idea was supported by another German astronomer, Franz von Paula Gruithuisen (originator of the impact theory of crater formation), who announced in 1822 that he had discovered a real 'lunar city' on the borders of the Sinus Medii, not far from the centre of the disk.

Gruithuisen was a keen-eyed observer who did much excellent work, but unfortunately his vivid imagination tended to bring discredit upon him even in his own lifetime. His 'lunar city' was a case in point. He described it as 'a collection of dark gigantic ramparts ... extending about 23 miles either way, and arranged on either side of a principal rampart down the centre ... a work of art'. Actually, his 'dark gigantic ramparts' turn out to be no more than low, haphazard ridges. Two of them are vaguely parallel for some distance, but there is nothing in the least like an artificial structure, and in any case the ridges are so low that they are difficult to see at all except when near the terminator. There can be no question of any change here, as Schröter, years before Gruithuisen, and Mädler, ten years afterwards, drew the region just as it is today.

Beer and Mädler showed that the Moon is definitely unable to support higher life-forms, and after the publication of their great book, in 1838, the moon-men were more or less handed over to the story-tellers, who used them to the full. Bug-eyed monsters, however, are relative newcomers to the literary scene (H. G. Wells was mainly responsible for them), and up to the time of Herschel it was thought more likely that the *Selenites* were humanoid.

Even so, the people of a hundred and forty years ago were quite ready to believe in bizarre life-forms. This led to the famous Lunar Hoax, the biggest scientific practical joke of all time apart possibly from the Piltdown Man.

Sir William Herschel had explored the northern skies with his great telescopes, discovering vast numbers of double stars, clusters and nebulæ, and probing the depths of space as no man before him had ever done. However, the southernmost stars, which never rise over England, remained comparatively unknown. Catalogues of the brighter ones were drawn up from time to time, but by the nineteenth century it had become clear that there was an urgent need for a more detailed survey.

Fittingly enough the task was undertaken by William Herschel's son, John.

On 13 November, 1833, Sir John Herschel set out for the Cape of Good Hope, taking with him a large telescope (you can see it today at Flamsteed House, the old Greenwich Observatory). Herschel stayed at the Cape for four years, and when he finally left, in 1838, his work had been well done. It took him more than a decade to collect and sort all the observations.

Herschel did not mean to pay any particular attention to the Moon or planets, which can be seen just as well from the northern hemisphere as from the Cape. He was concerned with the stellar heavens, and there was more than enough to occupy him. However, Richard Locke, a graceless reporter of the New York *Sun*, had a bright idea. Herschel was on the other side of the world; communications in those days were slow and uncertain; who was there to check any statements that he might care to make?

Locke saw his chance, and took it. On 25 August, 1835, the *Sun* came out with a headline reading 'Great Astronomical Discoveries Lately Made by Sir John Herschel at the Cape of Good Hope', and an account of how Herschel had built a new telescope powerful enough to show the Moon in amazing detail. Locke's article was so cleverly worded that it sounded almost plausible. It was well known, he wrote, that the chief limitation of any telescope is that it cannot collect enough light for extreme magnification, but Herschel had overcome this by effecting 'a transfusion of artificial light through the focal point of vision' —in other words, by using the telescope to form an image, and then reinforcing the image by means of a light-source in the observatory itself!

The way was open, and the *Sun* kept up the good work for the next six days. The lunar scenery was varied and colourful: 'A lofty chain of obelisk-shaped or very slender pyramids, standing in irregular groups, each composed of about thirty to forty spires, every one of which was perfectly square . . . they were of a faint lilac hue, and were very resplendent.' Next came animals, one of which was 'of a blueish lead colour, about the size of a goat, with a head and a beard like him, and a single horn . . . In elegance of symmetry it rivalled the antelope, and like him it seemed an agile sprightly creature running at great speed, and springing up from the green turf with all the unaccountable antics of a young lamb or kitten. This beautiful creature afforded us the most exquisite amusement.' On the next night the telescope was turned to 'a large branching river, abounding with lovely islands, and water-birds of numerous kinds . . . Near the upper extremity of one of these islands we obtained a glimpse of a strange amphibious creature of spherical form, which rolled with great velocity across the pebbly beach.' Botany was not neglected; in the crater Endymion 'Dr. Herschel has classified not less than 38 species of forest trees', but in the crater Cleomedes there seemed to be no living creature, except for 'a large white bird resembling a stork'. There were 'hills pinnacled with tall quartz crystals, of so rich a yellow and orange hue that we at first supposed them to be pointed flames of fire . . . a pure quartz rock, about three miles in circumference, towering in naked majesty from the blue deep; it glowed in the sun almost like a sapphire'. The brilliant Aristarchus was 'a volcanic crater, awfully rivalling our Mounts Etna and Vesuvius in the most terrible epochs of their reign . . . we could easily mark its illumination of the water over a circuit of sixty miles'.

The climax was reached on 28 August, with Locke's priceless account of lunar bat-men:

'Certainly they were like human beings . . . Having observed them for some minutes we introduced lens H.z, which brought them to an apparent proximity of eighty yards. They averaged four feet in height, were covered, except on the face, with short and glossy copper-coloured hair, lying snugly upon their backs. The face, which was of a yellowish flesh-colour, was a slight improvement upon that of the large orang-outang . . . The mouth, however, was very prominent, although somewhat relieved by a thick beard upon the lower jaw. These creatures were evidently engaged in conversation; their gesticulation, particularly the varied action of their hands and arms, appeared impassioned and emphatic. We hence inferred that they were rational beings.'

Locke was clever enough to bring in 'science' now and then. Sometimes a higher-powered eyepiece had to be used, sometimes conditions were unsuitable for observing, and sometimes it was necessary to turn up the hydro-oxygen burners to light up the faint image by the method of 'artificial transfusion'. On one occasion the astronomers forgot to cover up the main lens, so that when the Sun shone on it the lens acted as a vast burning-glass and set fire to the observatory.

The articles met with a mixed reception, but some eminent critics swallowed the bait hook, line and sinker. 'These new discoveries are both probable and plausible,' declared the New York *Times*, while the *New Yorker* thought that the observations 'had created a new era in astronomy and science generally'. A women's club in Massachusetts is said to have written to Herschel asking for his views on how to get in touch with the bat-men and convert them to Christianity, while even the Academy of Sciences in Paris held a debate when the news spread across to Europe, though it must be added that the French astronomers were highly suspicious!

The hoax was exposed by a rival paper within a few days, and the *Sun* itself confessed on 16 September. Even then, however, lingering doubts remained, and not for some months was the whole absurd business finally killed. Herschel apparently took the joke in good part when he heard about it, which was not until some time later.

It is easy to laugh; but can we afford to? Remember the 1938 panic, when a misleading broadcast of H. G. Wells' *War of the Worlds* led some people in the United States to believe that the Earth was being attacked by monsters from Mars; and even more recently, alarm and despondency was spread in South America by a radio announcer who became bored with the lack of news, and told his listeners that the Moon was about to fall upon the Earth (I understand that the broadcasting company subsequently dispensed with his services). Then, too, there are flying saucers. I do not propose to enter into a discussion about space crockery, but I cannot resist mentioning the occasion in 1958 when I interviewed the late Mr. George Adamski, co-author of the classic UFO book *Flying Saucers Have Landed*, on the B.B.C. television programme 'Panorama'. Mr. Adamski told me that he had been round the Moon, and had seen some dog-like creatures running about on the far side. Strangely enough, these interesting animals have not been confirmed by the astronauts who have since been round the Moon to see for themselves.*

Although the idea of intelligent Moon-men died over a century ago, animals and plants were still considered possible. In fact, the last great advocate of relatively advanced life on the Moon was W. H. Pickering, author of the 1904 photographic atlas as well as many papers about all branches of lunar study.

Between 1919 and 1924 Pickering, observing from the clear climate of Jamaica, carried out a detailed study of the noble crater Eratosthenes, which lies at the southern end of the Apennine range. He found some strange dark patches which seemed to show regular variations over each lunation, and although he was sure that tracts of vegetation existed on the Moon he thought it more likely that the Eratosthenes patches, which—he said—moved about and did not merely spread and shrink, were due to swarms of insects.

This startling idea was put forward in Pickering's final paper on the subject, published in 1924. He pointed out that a lunar astronomer of a century earlier would have seen similar moving patches on the plains of North America, due to herds of buffalo, and that the Eratosthenes patches were of about this size; on the other hand they moved more slowly—only a few feet per minute—and it was therefore reasonable to suppose that the individual creatures making them up were smaller than buffalo. Although insects were considered the most likely answer, Pickering's paper contains the following remarkable paragraph:

'While this suggestion of a round of lunar life may seem a little fanciful, and the evidence upon which it is founded frail, yet it is based strictly on the analogy of the migration of the fur-bearing seals of the Pribiloff Islands . . . The distance involved is about twenty miles, and is completed in twelve days. This involves an average speed of about six feet a minute, which, as we have seen, implies small animals.'

Pickering's idea was that the creatures, animal or insect,

* Mr. Adamski also told me that the inhabitants of Saturn play a kind of table tennis. I was once a County table tennis player myself, and I would dearly like to play against a Saturnian, but so far it has not been possible to arrange a fixture.

travelled regularly between their breeding-grounds and the dark 'vegetation' tracts nearby. His reputation ensured that due attention would be paid to his theory, and he clung firmly to his views right up to the end of his life in 1938. Few people agreed with him, but it was not until the coming of the space-ships that the concept of any kind of lunar life was completely ruled out. Even afterwards I can recall a paper by J. J. Gilvarry, who believed that the lunar maria used to be seas of water, so that the rocks would be of sedimentary type and the colours of the maria due to the remains of long-dead marine organisms.

The end came in July 1969, when Neil Armstrong and Edwin Aldrin came back from the Moon with their priceless lunar samples. Every precaution had been taken, and on their return the astronauts—including the third member of the team, Michael Collins—were placed in strict quarantine, just in case they had brought back any harmful lunar contamination. Only when the samples had been analyzed and found to be utterly sterile, was the danger regarded as non-existent, and quarantining was abandoned only with the later Apollo missions.

By now we have samples from various parts of the Moon, collected by the Apollo teams and by unmanned Russian probes, and the results are always the same. There is no trace of life, either past or present, and we can be certain that the Moon has been sterile throughout its long history. Primitive organisms of any kind are as unreal as Pickering's insects or Locke's bat-men. It has been left to men of our own world to bring life at last to the barren landscapes of the Moon.

Chapter Eleven

ECLIPSES OF THE MOON

A TOTAL ECLIPSE of the Sun is the grandest sight in all nature. As the brilliant solar disk is covered up, the corona flashes into view, and the effect is breathtaking. Unfortunately there are many people who have never seen a total eclipse of the Sun, because one has to be in just the right place at just the right time, and even then there are clouds to be considered. I count myself lucky to have seen four totalities, three of them under good conditions.

To recapitulate for a moment: a solar eclipse is caused when the new moon passes directly in front of the Sun. Because the Moon is only just large enough to blot out the Sun completely, the belt of totality can never be more than 169 miles wide; to either side of this narrow zone the eclipse is only partial, and the corona cannot be seen. For instance, the eclipse of 30 June, 1973 was total over parts of Africa, but from the south coast of England nothing was seen except for a small 'bite' out of the Sun, and over the rest of Britain there was no eclipse at all. The last total solar eclipse visible from anywhere in England was that of 1927. The next will not be until 30 August, 1999, when the line of totality will cross Cornwall. (Should you wish to see the eclipse in its full glory, I advise you to book your hotel accommodation early.)

With an eclipse of the Moon, the whole situation is different (Fig. 36). The eclipse is visible from a complete hemisphere of the Earth, because it is due to the Earth's shadow and not to any solid body blocking out the lunar disk. If a lunar eclipse is due, and the Moon happens to be above the horizon from your observing site, you will see it.

The result is, of course, that from any particular point on the surface of the Earth eclipses of the Moon are much commoner than eclipses of the Sun. On the other hand they are much less spectacular, and not nearly so important. All that really happens, from the viewpoint of the casual observer, is that the Moon becomes dim and changes colour during its passage through the shadow.

135

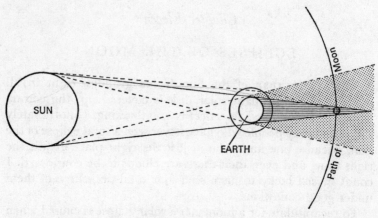

Fig. 36. A lunar eclipse

As Fig. 37 shows, the principal cone of shadow cast by the Earth, known as the umbra, is very long. Its average length is 850,000 miles, which is more than three times the distance of the Moon from the Earth, so that at the mean distance of the Moon (239,000 miles) the cone has a diameter of about 5,700 miles. In the second diagram I have tried to show this to a more accurate scale. In any case, it is obvious that the umbra is extensive enough to cover the Moon completely, and totality may last for as much as an hour and three-quarters, remembering that in the course of an hour the Moon moves across the sky by an amount which is slightly greater than its own apparent diameter.

Every scrap of direct sunlight is cut off from the Moon as soon as it passes into the umbra, and it might be thought that the Moon would simply vanish. This does not happen, because a certain amount of sunlight is refracted on to its surface by the Earth's atmosphere. In the diagram, one of these refracted rays is shown by the dotted line, and clearly it strikes the Moon, even though the Moon itself is directly behind the Earth. Instead of disappearing completely, the Moon becomes dim, and often shows strange and beautiful colour effects. Lunar eclipses may not be important, but they are undeniably lovely. During totality the effects of stars shining out near the dimmed Moon are particularly impressive.

Because the Sun is a disk, and not a mere point of light, the umbra of the Earth's shadow is bordered by what is termed the penumbra. This, too, causes a dimming of the Moon, but the effect is not nearly so marked. Since the Moon has to pass through the penumbra before it enters the umbra, there is a slight falling-off of light to one side of the Moon before the main eclipse begins. Many books claim that the penumbra is undetectable except by a skilled observer under good conditions, but I disagree strongly; I have never had the slightest difficulty in seeing the penumbra well upon schedule.

Not all lunar eclipses are total. Sometimes they are partial, and on other occasions penumbral, when the Moon avoids the main shadow-cone altogether. Obviously, an eclipse can take place only when full moon occurs near a node. On average, at least one lunar eclipse can be seen from any point on Earth each year, but not all of them are total. In 1975, for instance, there were two eclipses, both total; the first, on 25 May, was visible from the Americas, but from England the Moon was setting at about the time when the eclipse began. The second, on 18–19 November, was visible over Europe, Africa, western Asia and the north-eastern United States. The chances for 1976 were less good, as there was only one lunar eclipse— a partial one visible over Europe, Asia, Africa and Australasia. A list of lunar eclipses for the 1976–86 period is given in Appendix IV.

Eclipses may be predicted for many years ahead, because the movements of the Sun, Earth and Moon are so accurately

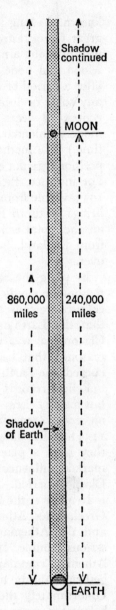

Fig. 37. Mean distance of Moon in relation to Earth's umbra

known. As long ago as 600 B.C. Thales of Miletus, first of the great Greek astronomers, was able to forecast eclipses fairly well by using a much rougher method. He knew that the Sun, Moon and node return to almost the same relative positions after a period of 18 years 10¼ days, the so-called Saros, so that any solar or lunar eclipse is apt to be followed by another eclipse 18 years 10¼ days later. (This is according to our modern calendar, and allows for five leap-years in the meantime.) The method is only approximate, because the relative positions are not exactly the same, but it is better than nothing. For instance, there was a total lunar eclipse on 29 January, 1953, visible from England. Adding one complete Saros period brings us to 10 February, 1971—and sure enough there was another total eclipse; but this time it was only partly visible from England, because the Moon set before the eclipse was over.

By using the Saros method, Thales was able to make some reasonable predictions, even though he had no idea of how eclipses are caused. In those early times it was not even known that the Earth is a globe. Yet by 450 B.C. Anaxagoras of Clazomenæ was well aware of what happened, and he also reasoned that because the Earth's shadow on the Moon is curved, the Earth itself must be spherical.

Eclipse records go back almost as far as written history itself, but most of them refer to the Sun, and the oldest lunar eclipse on record seems to have been that observed by the Chinese in 1135 B.C. Two later eclipses are worth mentioning here, because they have a place in history, and one of them even had a marked influence upon the whole sequence of events in the Classical period.

In 413 B.C. the Peloponnesian War was raging; the two chief Greek states, Athens and Sparta, were fighting for supremacy, and the Athenian army which had invaded Sicily was in serious trouble. In fact things were so bad that Nicias, the Athenian commander, decided to evacuate the island altogether, and if he had done so at once all might have been well. Unfortunately there was a total lunar eclipse on the night before the evacuation was due to begin, and Nicias believed that it had been sent as a warning. The astrologers were called in, and advised that the army should stay where it was 'for thrice nine days'. Nothing could have suited Gylippus, the enemy commander, better. He attacked the waiting Athenian fleet, destroyed most of it and blockaded the rest in harbour. The trapped Athenian army was utterly wiped out; Nicias lost his life, and eight years later Athens lay at the mercy of Sparta.

The story of the 1504 eclipse is not only more modern, but also more cheerful. At that time Christopher Columbus was in the island of Jamaica, and difficulties arose when the local inhabitants refused to supply him and his men with food. Unlike Nicias, Columbus knew a great deal about lunar eclipses, and he remembered that one was due on the first of March, a day or two ahead. He therefore told the Jamaicans that unless they mended their ways he would make the Moon 'change her colour, and lose her light'. The eclipse duly took place, so alarming the natives that they immediately raised Columbus to the rank of a god. No further trouble was experienced with food supplies!

The effect on the Jamaicans would have been even greater if the Moon had disappeared completely, as has been known to happen. There were two total eclipses in 1620, the first of which was watched by Kepler, and each time it seems that the Moon became utterly invisible; Hevelius noted the same thing in 1642, while in 1761 the Swedish astronomer Per Wargentin (after whom the famous lunar plateau is named) observed an eclipse in which the Moon vanished so thoroughly that it could not be found even with a telescope, though nearby faint stars shone out quite normally. Beer and Mädler saw a very dark eclipse in 1816, and in 1884 the shadowed Moon could only just be made out. On the other hand, the eclipse of 1848 was so bright that it was hard to tell that an eclipse was in progress at all, apart from the fact that the Moon turned a curious shade of blood-red.

I have seen a good many lunar eclipses by now, and no two are alike. For instance, on 29 January, 1953 the predominant colour was coppery-pink, with some glorious bluish hues and also what can only be termed flame-colour; at the partial eclipse of 15 July, 1954 there were reds and blues, though not so vividly, and on 24 March, 1959 the shadow was rusty red. The eclipse of 30 December, 1963 was unusually dark, and so was that of 10 December, 1973, though it is true that proper estimates were not easy on that occasion because only 10 per cent. of the Moon was covered by the umbra.

Obviously, conditions in the Earth's atmosphere have much to do with these variations, because all the light reaching the eclipsed Moon has to be refracted through our air. Various correlations have been found. The tremendous volcanic explosion of Krakatoa in 1883 scattered so much dust in the upper atmosphere that its effects were traceable for months afterwards, and probably caused the darkness of the 1884 eclipse. Other dark eclipses can be similarly linked; that of 1902 with the eruption of Mont Pelée, that of 1950 with vast forest fires raging in Canada, that of 1963 with the outbreak at Mount Agung in the East Indies, and so on. But this may not be the whole story.

The French astronomer A. Danjon (whom I knew well; he died some years ago) introduced an 'eclipse scale' which is distinctly useful. On this system, 0 indicates a very dark eclipse, with the Moon almost invisible near mid-totality: 1, dark eclipse with greys and browns, and with details on the surface barely identifiable; 2, deep red or rusty, with the outer edge of the umbra relatively bright; 3, brick-red, with a bright or yellow rim to the shadow; 4, very bright, orange or coppery-red, with a bright bluish shadow rim. Danjon also tried to link the brightness of a lunar eclipse with conditions on the Sun. It has long been known that the Sun has a reasonably regular cycle of activity, with spot-maxima occurring every eleven years (in 1957–58, for instance, and again in 1969), while near minimum activity the solar disk may remain spotless for many days consecutively. According to Danjon, eclipses of the Moon are dark for the two years after solar maximum; they then become brighter until the seventh or eighth year, when they reach 4 on his scale; subsequently there is an abrupt decrease. If this relationship is valid, the eclipse of April 1977 should be bright. We must wait and see.

There have been recent suggestions that there is more to it than simple refraction, because conditions in the Earth's upper air may well be affected by what is happening in the Sun, and this would in turn affect the darkness of lunar eclipses. Another idea was that the luminescence of the lunar rocks themselves could be involved, but it has now been shown that luminescent effects are too slight to be detectable.

To the selenographer, the most important fact about a lunar eclipse is that there is an abrupt cut-off of sunlight. Since the Moon is without atmosphere, and the surface rocks are very poor at retaining heat, a sudden wave of cold sweeps across the Moon. During the 1939 eclipse, Pettit and Nicolson at the Mount Wilson Observatory found that the temperature dropped from $+160°$ F. to $-110°$ F. in only an hour, and this sort of result has been confirmed since. (The temperatures measured at radio wavelengths do not show the same variation, because here we are dealing with regions slightly below the Moon's surface, and the outer rocks are excellent insulators.)

Yet not all parts of the Moon cool down at the same rate, and during the past few years there have been some important researches carried out in the infra-red, mainly by the American astronomers J. M. Saari and R. W. Shorthill. Infra-red, of course, is long-wavelength radiation, which does not affect the eye; many people are only too familiar with the infra-red lamps used in hospitals to radiate heat. Bodies in the sky radiate in the infra-red—it would be most surprising if they did not!—and the Moon is no exception. Nowadays, infra-red astronomy is becoming more and more important in the scientific scene.

Using infra-red techniques, Saari and Shorthill have found that during an eclipse there are some regions which cool down relatively slowly, and these regions have become known as 'hot spots'. Actually, the term is misleading; all it means is that when the Moon is in the shadow of the Earth, these spots are less chilly than their surroundings. Over four hundred such areas have been located, some of them associated with major craters. The best example is Tycho, centre of the bright ray system, in the southern part of the Moon.

There is no suggestion of internal heat being responsible, and in any case the effect is not confined to eclipses. The 'hot spots' are warmer than their surroundings during the long lunar nights, though they are apparently cooler than their surroundings during the daytime. The exact reasons for this behaviour are still not known. The chemical or mineralogical make-up of the outer surface layer may play a part; also, it may well be that there are exceptional numbers of rocks and boulders lying around, which will tend to stabilize the temperature (i.e. by lunar standards). At least there seems to be a general pattern, inasmuch as two of the main 'hot spots' are Tycho and Copernicus, the most important ray-craters on the Moon. Aristarchus is another; so are smaller bright craters and minor ray-centres, though not all the 'hot spots' are associated with rays.

Whether any visible effects can be tracked down in surface features is much more dubious, and our ideas have changed markedly during the past half-century. W. H. Pickering (of lunar insects fame!) believed that the sudden onset of cold could cause precipitation of snow or hoar-frost. In particular, he concentrated upon the little formation Linné in the Mare Serenitatis, about which I shall have more to say when talking about possible long-term changes on the Moon. At present, Linné is a craterlet surrounded by a white patch or nimbus. Pickering believed the white nimbus to be due to frost, and he expected that the size of the patch would increase perceptibly during an eclipse.

Pickering and A. E. Douglass in America, and S. A. Saunder in England, made careful measurements of Linné during successive eclipses, and came to the conclusion that there was definite growth in the whiteness; after the end of the eclipse, when the surface was heated up again, the patch reverted to its normal size. Frankly, I am profoundly sceptical. I have made measurements at many eclipses with totally negative results, and I have been equally unsuccessful with other white patches of the same basic type. Nowadays, nobody believes in the possibility of hoar-frost, and any change in Linné or similar formations during an eclipse would be hard to explain. Neither is there any conclusive proof of any eclipse-induced changes in the dark-floored plains such as Grimaldi and Plato. Variations are worth looking for at suitable eclipses, but in my view there is little chance of any positive result.

There is no glare from the full moon during an eclipse, and occultations of stars can be well seen. Unfortunately, bright stars are seldom in the right place at the right time, and the only recorded instance of a bright planet (Jupiter) being occulted by a totally-eclipsed Moon dates back as far as the year 755. In the days when it was still thought possible that the lunar atmosphere might be dense enough to produce meteors, careful searches were carried out during eclipses, when conditions for the detection of short luminous trails would be at their best; but the searches met with no success—because, as we now know, the Moon's atmosphere is far too tenuous to heat infalling particles to luminosity.

A lunar eclipse is not so exciting as a total eclipse of the Sun; there are no prominences or coronal rays, and everything happens much more slowly. Yet it cannot be denied that the passage of the Moon through the dark cone of shadow thrown by our own world has a quiet fascination all its own.

Chapter Twelve

THE WAY TO THE MOON

THE IDEA OF TRAVEL to the Moon is by no means new. In the second century A.D. a Greek satirist, by name Lucian, wrote a story called the *True History* which described a lunar voyage carried out quite involuntarily. According to Lucian, a ship carrying a full crew through the Straits of Gibraltar was caught in a waterspout, and was hurled upward for seven days and seven nights until it landed on the Moon. Lucian did not mean to be taken seriously—he admitted that his 'true history' was nothing but lies from beginning to end—but even in those far-off times it was known that the Moon, like the Earth, had a surface covered with mountains, valleys and plains.

Various other extravagant suggestions were made during the period following the invention of the telescope. In Kepler's classic *Somnium* the hero was carried moonward by demon power. Four years later, in 1638, came *The Man in the Moone*, written by an English bishop, Francis Godwin, in which the astronaut was flown to the Moon on a raft pulled along by wild geese—to be greeted by a race of giants talking in a language so musical that it could be written down only in note form. Lunar society, wrote Godwin, was distinctly puritan. Any Moon-child showing signs of latent wickedness was promptly dispatched to the Earth, where there is already so much evil that a little more will not matter. (Today, looking round the world of A.D. 1976, who can say that Godwin was wide of the mark?)

Godwin, like Lucian, was writing with his tongue very much in his cheek, but another English bishop of the same period, John Wilkins, was not. In his *The Discovery of a World in the Moone*, Wilkins maintained that the Moon is inhabited, and suggested that the British Government should annex it for the nation. The interesting thing about this book is that the author meant it to be taken very seriously indeed, and by the standards of his time he was no eccentric; he later became Secretary of the Royal Society.

Passing over Cyrano de Bergerac's *Voyages to the Moon and Sun*, in which the space-travel methods ranged from sucked-up dew to exploding fire-crackers, we must come forward to 1865, which is an important date in lunar history inasmuch as it saw the publication of one of the most famous science-fiction stories of all time. *From the Earth to the Moon*, by Jules Verne (followed a few years later by its sequel, *Round the Moon*) introduced the idea of a space-gun, in which the travellers were sent to their target inside a projectile fired from a huge cannon. Verne was not himself a scientist, but he believed in checking his facts, and his novel makes splendid reading even today. Obviously, the whole concept of the space-gun is out-moded; but Verne chose the correct escape velocity (7 miles per second), his fictional launching-ground was not far from the modern Cape Canaveral, and the giant telescope he described on Long's Peak echoes the real Hale 200-inch reflector on Palomar Mountain. It is, I feel, significant that when the Russians sent their first camera-carrying vehicle on a trip round the Moon, in 1959, one of the far-side craters that they discovered was promptly named in honour of Jules Verne.

This is not a history of space research, and in any case I have dealt with the story in more detail elsewhere.* So I will do no more than say that the only valid method of sending vehicles to the Moon is by rocket power, since this involves the principle of reaction—'every action has an equal and opposite reaction'—and there is no need for a surrounding atmosphere, as there is with an ordinary aircraft. In a Guy Fawkes rocket, for instance, the hot gas rushing out of the exhaust propels the tube of the rocket in the opposite direction; the rocket kicks against itself, so to speak, and is at its best in vacuum, where there is no resisting atmosphere to be pushed out of the way. The principle of a modern space-vehicle is the same, even though the solid gunpowder of the firework is replaced by a tremendously complicated rocket motor powered by liquid propellants. One man who worked out the main theory in considerable detail, even before the end of the nineteenth century, was the Russian schoolmaster Konstantin Eduardovich Tsiolkovskii, who pub-lished some articles about it in obscure journals which caused absolutely no comment—for the simple reason that almost

* In my book *Space* (Lutterworth Press, 1973).

nobody knew about them. It was only toward the end of his life that Tsiolkovskii achieved fame. He never built a rocket, and was in no position to try; but I recommend you to read his novel *Beyond the Planet Earth*, written originally in 1896. As a story, and as a literary exercise, it can only be described as atrocious, but as a forecast it was decades ahead of its time.

Tsiolkovskii knew that liquid fuels would have to be used instead of the weak, uncontrollable solids of the gunpowder variety. In 1926 Robert Hutchings Goddard, in America, actually fired the first liquid-propelled rocket in history, managing an altitude of 184 feet and a top speed of 60 m.p.h. Again there was no general comment, because Goddard was not interested in publicity; but between that modest beginning and the start of the war, thirteen years later, there were spectacular developments, mainly in Germany. An initially amateur team was taken over lock, stock and barrel by the German Government, and on Hitler's orders a research base was set up on the Baltic island of Peenemünde specifically to develop rockets for use in war. Out of this research came the V.2, which Londoners will remember with anything but affection. Yet even though the V.2 was built for destruction, it was the direct ancestor of the Apollo vehicles which have sent men to the Moon; and one of those present in Mission Control, Houston, when Armstrong and Aldrin touched down on the lunar surface, was Wernher von Braun, late of Peenemünde.

Following the coming of peace and the start of the cold war, the centre of interest shifted to America. Rockets were improved year by year, and even the most diehard opponents of space-travel began to realize that the Moon was coming within reach. Yet before long the initiative passed to the Soviet Union, and it is my firm view that the real start of the Space Age can be fixed as 4 October, 1957, when the Russians launched their first artificial satellite—Sputnik 1, which sped blithely round the world sending back its 'Bleep! bleep!' radio signals which sounded, to some people, faintly derisory. In the following year the American team led by von Braun managed to send up a satellite, and also did their best to dispatch vehicles to the Moon. Four launches ended in failure, and again the Russians achieved a notable 'first'.

Strictly speaking there were three 'firsts', all in 1959 (Fig. 38).

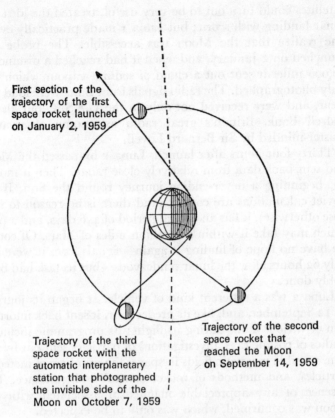

First section of the trajectory of the first space rocket launched on January 2, 1959

Trajectory of the third space rocket with the automatic interplanetary station that photographed the invisible side of the Moon on October 7, 1959

Trajectory of the second space rocket that reached the Moon on September 14, 1959

Fig. 38. Paths of the first Lunas

In January the probe Luna 1 passed within 4,660 miles of the Moon, proving, among other things, that there is no appreciable lunar magnetic field. In September, Luna 2 made a crash-landing on the lunar surface, thereby forging the first direct link between our world and another; and on 4 October— exactly two years after the ascent of Sputnik 1—Luna 3 started on a journey which took it right round the Moon, giving us our first positive information about those parts of the surface which are always turned away from the Earth.

It is, I think, justifiable to say a little more about these 1959 vehicles, because they were so significant. Even after Sputnik 1 there were still people who, while conceding that artificial satellites could turn out to be very useful, treated the idea of a lunar landing with scorn; but Luna 1 made practically everyone realize that the Moon was accessible. The probe was launched on 2 January, and when it had reached a distance of 70,000 miles it sent out a cloud of sodium vapour which was duly photographed. The radio signals from Luna were loud and clear, and were received not only in the U.S.S.R. but also at Jodrell Bank, Britain's great radio astronomy observatory master-minded by Sir Bernard Lovell.

Thirty-four hours after launch, Luna 1 by-passed the Moon and sent back data from relatively close range. Then it moved on, beginning a never-ending journey round the Sun. If the Soviet calculations are correct (and there is no reason to suppose otherwise), it has an orbital period of 446 days, and a path which may take it within six million miles of Mars. Of course we have no hope of finding it again—signals from it were lost only 62 hours after the lunar rendezvous—but its task had been nobly done.

Luna 2 was a different kind of vehicle. It began its journey on 12 September, and, like its predecessor, it sent back information of various kinds during its flight; its programme included studies of cosmic rays, investigations of particles sent out by the Sun, possible magnetic fields in space, the numbers of meteoric particles, and methods of rocket control and guidance. The absence of any appreciable magnetism associated with the Moon was confirmed, which was only to be expected.

Before hitting the Moon, the last stage of the compound rocket was separated from the container carrying the scientific equipment. Presumably, no effort was made to bring either component down gently; this, remember, was in the Neolithic period of space research! The predicted time of impact was 21 hours 0 minutes G.M.T. on 13 September, and observers all over the world were at their telescopes, hoping to see some sign of a flash as Luna 2 reached the end of its journey. The results from the Jodrell Bank 250-foot radio telescope showed that as it neared its target the probe was still sending back loud, clear signals. At 21 hours 2 minutes 23 seconds the signals ceased abruptly. This, then, was the moment of impact.

I was using the 12½-inch reflector at my observatory, then at East Grinstead in Sussex, and since the Russians had given an approximate landing-area I risked a fairly high magnification. Suddenly I saw, or thought I saw, a tiny spark, so faint as to be at the limit of visibility. The time was 21 hours 2 minutes 23 seconds—though I did not then know what was happening, and all I had to work on was the original estimate. When I checked up and found that H. P. Wilkins, using his 15-inch reflector in Kent, had recorded the same sort of spark at the same moment and in the same position, I wondered whether I had in fact seen the landing of the carrier-rocket. On the whole, however, I am more inclined to dismiss it as a trick of the eye; various other observers in Europe recorded flashes in different places, and there was no general agreement. The whole episode was quite unimportant, though I admit I am glad to have been on watch at the time when a man-made vehicle made our first contact with the Moon.*

Next, in the following month, came Luna 3, the first probe to send back really vital information about the Moon's surface. It had one main task: to go round the Moon, and tell us what the enigmatical far side was really like.

The question of 'What lies on the other side of the Moon?' had been tantalizing astronomers for many centuries. It was so infuriating not to know; the Moon, after all, is extremely close to us, but until October 1959 our knowledge of its surface was restricted to the 59 per cent. which can be studied from Earth.

The various librations have been described in Chapter 5, but another chart (Fig. 39) will help to show their effects. In the left-hand drawing, the large crater Gerard is shown, together with the lunar limb at mean libration. The same area is shown in the right-hand sketch, but this time with the most favourable libration possible for that region; Gerard is much better placed, and new details, actually on the far hemisphere, have come into view. Yet one inescapable fact remained: 41 per cent. of the Moon was absolutely inaccessible to observation, and nobody really knew what it was like.

Speculation was not lacking, and some of the ideas put

* I must reluctantly gloss over other reports, such as one from an earnest lady who telephoned me saying that she had been watching the impact with binoculars, and had seen the Moon split in half. Incidentally, the generic term for the probes in common use at that time was 'Lunik'; it became 'Luna' later on.

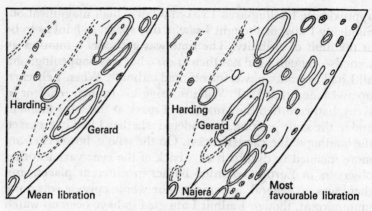

Fig. 39. The Gerard area

forward were fascinating. There was, for instance, a theory due to a famous last-century Danish mathematician, Hansen, who was busy studying the movements of the Moon when he found some discrepancies which he could not explain. They led him on to suggest that the Moon is not uniform in density, but has one hemisphere slightly more massive than the other. This lopsidedness would shift the centre of gravity some way from the centre of figure, and he worked out that it was 33 miles further from the Earth. He concluded that all the Moon's atmosphere and water had been drawn round to the far side, which might well be inhabited!

Not many people agreed, and when Hansen's discrepancies were satisfactorily explained without recourse to lop-sidedness the whole theory was relegated to the scientific scrap-heap. (It is true that the centre of mass is slightly away from the centre of figure, as the space-probes have shown; but the amount is less than two miles—a point to which I will return later.) The situation was summed up rather neatly by a famous poem which was, I believe, written by a housemaid with literary aspirations, and which has been handed down to posterity. There seem to be several versions of it, and I have chosen the most-quoted one:

> O Moon, lovely Moon with the beautiful face,
> Careering throughout the bound'ries of space,
> Whenever I see you, I think in my mind
> Shall I ever, O ever, behold thy behind?

150

All that could really be said was that the far side was likely
to be just as barren, just as hostile and just as airless as the side
which we have always known. When it came to the question of
the arrangement of craters and mountains, or the frequency of
maria, astronomers could do little more than make intelligent
guesses.

One interesting investigation—now, of course, of historical
value only—was carried out during the last century by the
American geologist, N. S. Shaler. Shaler looked at the ray-
craters on the visible side, and realized that their positions
could be plotted from their rays alone. He therefore started to
examine the limb regions to see whether he could trace any
rays which came from the far side, and which might give a clue
as to the positions of the craters responsible for them. As he had
expected, there were a few, and eventually he plotted six
possible centres, all well on the hidden hemisphere. Unluckily
he mislaid his notebooks, and when he returned to the problem,
thirty years later, he could remember the rough positions of
only three of his ray-centres. Neither could he re-observe
the rays, since his eyesight was no longer sufficiently keen.
During the 1930s an English observer, E. F. Emley, obtained
results similar to Shaler's, and subsequently Wilkins used
Emley's observations, together with his own and some of mine,
to locate eight possible ray-centres on the far side of the Moon.
We now know that his results were reasonably accurate, but at
the time there seemed little prospect of finding out.

During the immediate post-war years, various observers
were doing their best to plot the libration regions near the very
edge of the Moon's disk. It was not easy, because everything
was horribly foreshortened, and it was difficult to tell a crater
from a ridge; also, one had to take advantage of the rare
occasions when the libration in any particular region was
favourable at the time of suitable solar illumination. All the
same, it was immensely fascinating, and there was always the
chance of making a discovery. I did so myself on one occasion,
when I happened upon a large walled plain right on the limb at
maximum libration—beyond the huge, ruined formation Otto
Struve (shown on Map 8 in the Appendix). I duly reported it,
and it was subsequently photographed; the name originally
given to it was Caramuel, though the name was later altered to

Einstein. I can claim no credit whatsoever for the discovery; I was merely looking in the right place at the right time. Then there was the case of the apparently small, mare-like feature on the limb, which Wilkins and I found during an evening's work at his observatory in Bexleyheath. We were not sure of its nature, but we wrote a paper about it and christened it 'Mare Orientalis'—the Eastern Sea (now amended to Mare Orientale). What we did not know, of course, was that it would prove to be a vast, amazingly complicated and important structure extending right on to the Moon's far side. I will always have something of a fatherly affection for it!

Orientale apart, no major sea extends on to the hidden 41 per cent. of the Moon, and in 1952 I dared to make some forecasts. I expected no large plains of the Mare Imbrium type, and I predicted that the crater arrangement would be more haphazard than on the familiar side. This was because I believed then, and I believe now, that the main lunar features are of internal origin, and that the Earth's gravitational pull has had a great deal to do with their distribution. My reasoning may or may not have been valid, but the forecast was correct. Unfortunately, I added that no proof could be expected for many years. Actually it took less than eight, and this brings me back to Luna 3.

Like its predecessors, Luna 3 was given a full research programme, but its main task was to pass beyond the Moon and photograph the far side (Fig. 40). The launcher was one of the conventional step-vehicles; it had to be powerful, because of the weight of its load. Without fuel, the upper stage of the rocket weighed about a ton and a half, while the weight of the 'station' itself amounted to nearly nine hundredweight— massive by 1959 standards.

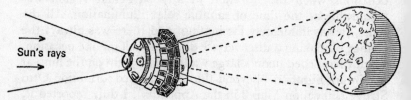

Sun's rays

Fig. 40. Position of Luna 3 during the photography of the far side of the Moon

All went well. By 4.30 G.M.T. on 7 October the rocket had passed by the Moon, and lay beyond it, at a distance of under 40,000 miles from the lunar surface. The photographic apparatus was switched on, and for the next forty minutes the pictures were taken. Two cameras were used, giving photographs on different scales. After the programme had been completed, the films were automatically developed and fixed ready for transmission back to Earth.

Delay was inevitable, because Luna 3 was still receding from us. It reached its apogee or furthest point on 10 October, when it was 292,000 miles away, and then started to swing in once more, reaching perigee (29,000 miles) on 18 October. It was then that the pictures were sent back. They were scanned by a miniature television camera, and the transmissions were picked up by the waiting Russians. Late on 24 October, the photographs were given to the world.*

Blurred and lacking in detail though they are by modern standards, the Luna 3 pictures represented a tremendous technical triumph. Several features on the far side showed up, notably a dark-floored walled plain which the Russians christened Tsiolkovskii in honour of the famous rocket pioneer. Another dark feature was named the Mare Moscoviense or Moscow Sea. Bright craters were also detectable, with some ray-systems. Inevitably there were errors in interpretation—a suspected high mountain chain, which was named the Soviet Range, later turned out to be non-existent—but it was an encouraging start. Evidently the Russians hoped to re-run the pictures later on, but contact with Luna 3 was lost abruptly and was never regained, so that we do not know what happened to it.

The next four years were less fruitful. At this stage the American lunar probe programme was frankly floundering; the vehicles either went out of control, exploded, or missed the Moon completely. The Russians had a similar failure with

---

* I shall not forget my first sight of them. At 10.15 p.m. I was just starting a live broadcast in my B.B.C. television series 'The Sky at Night', and the pictures came through direct on the screen, giving me no time at all to think out a suitable commentary. Luckily the Mare Crisium was shown clearly, though naturally from an unfamiliar angle, and I was able to get my bearings.

Luna 4 in April 1963; they may have been trying for a soft landing on the surface, but with no success. The next real achievement came with America's Ranger 7, which hit the Moon on 31 July, 1964. Inevitably it destroyed itself, but during the last minutes of its flight it sent back over four thousand high-quality photographs, and for the first time the surface of the Moon could be studied from really close range.

The impact point was in the Mare Nubium or Sea of Clouds, near the 36-mile, low-walled crater Guericke (Map 12 in the Appendix). (The region was promptly named the Mare Cognitum or Known Sea, though it does not seem that this name has found its way on to many maps.) Four cameras were used altogether, and were in operation for only just over a quarter of an hour, so that the area covered photographically was somewhat restricted—about the same as that of France. The last picture was transmitted only 0·19 of a second before impact, and showed a region of the Moon measuring 105 feet by 150 feet, with crater-pits down to a few inches in diameter. There were many small features, and some rounded depressions, almost without walls, which looked like collapse features. Early in 1965 two more Rangers were equally successful. No. 8 (20 February) landed in the Mare Tranquillitatis, and No. 9 (24 March) inside the walled plain Alphonsus, which was—and is—of special interest because it is thought to be one of the more active areas of the Moon.

The results from these three crash-landers were much the same, so I can deal with them jointly. Even the smooth-looking parts of the Moon turned out to be anything but level when seen from point-blank range; there were pits, hummocks and ridges everywhere. Some of the rills inside Alphonsus proved to be made up of chains of small craters which had run together, and the whole lunar surface was rock-strewn. This brings me on to another controversy which was very much to the fore in the early 1960s. Would the Moon's crust be firm enough to support the weight of a space-craft?

One man who had his doubts was Dr. Thomas Gold, then of the Royal Greenwich Observatory. In 1955 Gold, one of the world's leading mathematical astronomers, published a paper in which he claimed that the lunar maria, at least, were filled with soft dust—a treacherous ocean, kilometres deep, into which a probe would inevitably disappear. In his own words, the luckless craft 'would simply sink into the dust with all its gear'. Naturally, Gold's opinions carried a great deal of weight, and space-planners were decidedly apprehensive. To achieve a successful touch-down, only to lose the vehicle in a matter of seconds, would be most disappointing, but if Gold's theory were correct nothing much could be done.

Practical observers were not impressed. According to Gold the lunar dust would flow 'downhill', coming to rest in the lowest-lying parts of the Moon; of course the maria are at lower levels than the bright areas. Yet there are some craters on the maria whose floors are of exactly the same texture as the surface outside. Archimedes, on the Mare Imbrium, is one. To invade the crater, Gold's dust would have had to have climbed over the walls; in other words it would have had to have flowed uphill, which did not seem very likely. Also, dust would have dropped into the rills and produced dark floors, whereas in fact rill-bottoms are bright. There were other objections too, and I for one had no faith whatsoever in the alleged dust-layer; but there was only one test—send a probe to find out.

This is precisely what the Russians did, in February 1966. After several failures they launched Luna 9, which was slowed down by rocket braking while still well above the Moon, and dropped gently on to the surface near the edge of the Oceanus Procellarum, not very far from the dark-floored walled plain Grimaldi. Within a few minutes the first signals were being sent back from the lunar surface, and were received both in the Soviet Union and by Sir Bernard Lovell's team at Jodrell Bank. One fact emerged immediately. Luna 9 was standing on a hard layer, with no tendency to sink, so that Gold's dust theory was completely wrong. The scene was remarkably like that of a lava-field in Iceland or some similar place; there were various rocks and boulders strewn around, and the whole landscape was rough. Potential astronauts felt comforted. If a manned craft were to land, the Moon would at least refrain from swallowing it up.

As has so often happened, a Russian success was quickly followed by something comparable from America, and between June 1966 and January 1968 seven Surveyors were sent to make soft landings on the Moon. Five were successful; of the others,

No. 2 went out of control at the critical moment and crashed to destruction, while contact with No. 4 was lost before touchdown. During the same period the Soviet team dispatched Luna 13 (December 1966), which came down safely in the Oceanus Procellarum.

The results from these various vehicles were all much the same. The firmness of the lunar ground was confirmed, and many excellent photographs were sent back; even the first Surveyor managed more than 11,000, and contact with it was kept up for seven months.

Surveyor 3, launched on 17 April, 1967, was also aimed at the Oceanus Procellarum, some 230 miles south of Copernicus. Its programme was not confined to photography. It carried a sort of mechanical scoop, so that the texture and mechanical properties of the lunar 'soil' could be tested. I mention Surveyor 3 particularly since over two years later Charles Conrad and Alan Bean, from Apollo 12, landed so close to it that they were able to walk over and examine it; Surveyor was apparently undamaged, though of course, its power had long since failed, and the astronauts were able to hack pieces off to bring home for analysis.

The last three Surveyors were even more fruitful, since they carried what may be called chemical sets, and were able to confirm the long-held suspicion that the lunar surface is made up essentially of grey volcanic rock known to geologists as basalt. Numbers 5 and 6 landed in relatively smooth areas (the Mare Tranquillitatis and the Sinus Medii respectively), but Surveyor 7 touched down on the northern outer slopes of Tycho, so that it was the first probe to visit a highland area. Analyses showed that there was more aluminium but less iron than in the maria, though the composition of the material was still essentially basaltic.

Valuable though these soft-landers were, they were matched by the American Orbiters, of which there were five—all highly successful. The Orbiter vehicles were put into closed paths round the Moon, so that full photographic coverage could be achieved. So many pictures were obtained that even today many of them are stacked away to await analysis. Orbiter 1 was launched on 10 August, 1966. The fifth and last probe of the series began its journey on 1 August, 1967; and when it came to the end of its career the task of mapping the Moon, begun by Harriott and Galileo so long before, was to all intents and purposes complete. Practically the whole of the Moon was photographed; the only region left out was a small part of the south polar zone, which was, fortunately, covered by similar Russian orbiting vehicles. The far hemisphere was mapped as adequately as the region accessible from Earth, and this therefore seems the right moment to say a little more about the Moon's 'other side'.

The overall aspect is not the same as that of the familiar hemisphere. It has been said that the far side is all highland, which is a very fair description of it. Large maria are entirely absent, though it is true that the Mare Orientale extends well over the limit of the region observable from Earth, and the same is true of a few of the minor seas along the eastern limb of the disk as we see it (that is to say, beyond the Mare Crisium; see Map 1). The so-called Mare Moscoviense and the rather smaller Mare Ingenii, on the far side, are not genuine maria. On the other hand there are various large basins, with light-coloured floors not filled with Mare material. To anticipate Chapter 14 for a moment: it seems that the crust is thicker on the far side than on the Earth-turned side, which may explain why lunabase has not welled up through the crust to the same extent. The centre of gravity of the Moon really is displaced away from the geographical centre, and is further away from the Earth; but the difference is less than two miles, so that there is no suggestion of a return to Hansen's theory!

There are some splendid craters, plains and valleys on the far side. One particularly impressive crater-valley is associated with the large formation which has been named Schrödinger, only just out of view from Earth. As I predicted long ago (probably more by luck than by judgement) the arrangement of the craters is less regular, though the general laws of distribution still apply; it is usually a smaller crater which breaks into a larger, not vice versa.

Tsiolkovskii is probably the most imposing object on the far side, if we except the Mare Orientale. The main characteristic of Tsiolkovskii is the darkness of its floor; in many photographs it gives the impression of being shadow-filled, though the real cause of the darkness is purely the colour of the interior itself.

There seems little doubt that we are seeing a lake of dried lava, from which a central peak rises. Tsiolkovskii intrudes into a larger, less regular basin, Fermi, whose floor is of the usual light hue.

One minor controversy has centred around the naming of the various features on the far side. Since the Russians were first in the field, with Luna 3, they felt entitled to allot what names they liked to the objects they managed to identify, and nobody was in a position to argue. Most of these names are still recognized. Later the Americans joined in, and the whole matter was handed over to the International Astronomical Union, the controlling body of world astronomy. The system of naming lunar features after famous personalities has been followed, and I hope that the present nomenclature is satisfactory. Because nobody can actually see the far side of the Moon without going up in a space-ship, the maps I have given in this book are much less detailed than those of the near side, and I have made no attempt to put in all the names; I have confined myself to the major features, and have drawn the charts to a much smaller scale. Still, I hope they will serve to give the general impression.

Obviously the maps have been culled from the numerous space-probe pictures (Lunas, Orbiters and Apollos), and the views look unfamiliar to earth-bound lunar observers such as myself. It is strange to see what looks like a full moon but with—say—the Mare Crisium in the centre of the disk. Then, too, the craters which we Earth-dwellers see as foreshortened are shown in their true guise; look for instance at Clavius as photographed 'from above'—it looks (and is) circular. Then, too, there is the huge walled plain Wilhelm Humboldt, which from Earth appears as a long, narrow ellipse close to the limb. Probe pictures show its floor clearly, and reveal a complicated system of rills.

Quite apart from the photographic results, the Orbiters (and the later Apollos) sent back data of all kinds, including measurements of the shape of the Moon. They have found that the axis pointing to the Earth is about a kilometre longer than the other axis, so that the Moon is very slightly egg-shaped. Also, the movements of the probes have led to the discovery of what are called mascons, whose origin is still very much a matter for debate.

The first announcement of mascons came in 1968, with some work by the American astronomers P. Muller and W. L. Sjogren, who were studying the movements of Orbiter 5. They kept a careful check on Orbiter during eighty consecutive revolutions round the Moon, each taking 3 hours 11 minutes, and found that the velocity in orbit was not constant. There were times when there was a slight speeding-up, followed by a slowing-down to the original rate. The effect was very small indeed, but it could be measured.

If a probe passes over a region where the surface material of the Moon is unusually dense, then the probe will be pulled along, so to speak, and it will speed up. Muller and Sjogren found that this happened over the same areas during each revolution round the Moon, and this led them to assume the existence of dense masses below the lunar crust. It was soon found that there was a strong correlation with the lunar seas, which are depressed below the upland level—the Mare Crisium and the Mare Smythii, for instance, by as much as two and a half miles. The term mascon (a convenient abbreviation for *mas*s *con*centration) came into general use, and explanations were sought.

According to one early suggestion, a mascon is nothing more nor less than a meteorite which hit the Moon thousands of millions of years ago, coming to rest far underground and blasting out a basin which was subsequently filled with lava to become a lunar 'sea'. However, a meteorite capable of affecting the motion of an Orbiter vehicle would have to be either improbably large or else improbably dense, and I think it is true to say that the meteorite theory has now been abandoned by practically everybody. A mascon is more likely to be a flat or disk-shaped body made of volcanic rock, and lying close to the Moon's surface.

Mascons have been found below the regular seas (Imbrium, Crisium, Smythii, Serenitatis, Humorum, Nectaris, Humboldtianum, Orientale and Sinus Æstuum) and also below a few dark-floored walled plains, of which Grimaldi is probably the best example. It seems, then, that the extra mass (producing what is termed a positive gravity anomaly—that is to say, a greater pull than average) is associated with the material which filled the Mare basins more than 3,000 million years ago. It also supports my idea that features such as Grimaldi are of the same type as the maria. On the other hand, craters of the Copernicus type tend to have negative gravity anomalies, and the same is true of the basins which are not filled with mare material. Most of these unfilled basins lie on the far side of the Moon.

From what I have said here, it is obvious that our knowledge of the Moon was increased tremendously between 1959, when Luna 1 made its flight, and 1967–68, when the Orbiter series came to an end. But there is a vital difference between exploring with automatic probes, and going to see for oneself; and even while Orbiter 5 was still transmitting, the programme of manned travel was well under way. America had announced her intentions, and had started to put them into practice. The real attack on the Moon was about to begin.

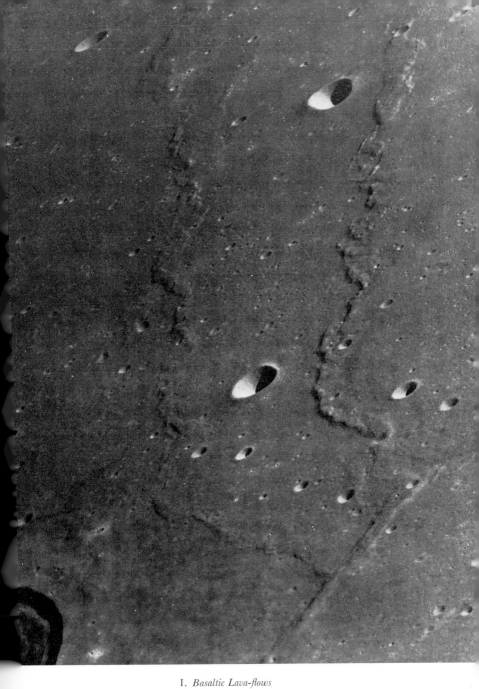

I. *Basaltic Lava-flows*

Lava-flows on the basaltic surface of the Mare Tranquillitatis. Photograph taken from
Apollo 10.

II. *Tycho*

Tycho; a sunrise view, taken from the 80 mm camera on Orbiter 5. Surveyor 7 landed north of the main wall—that is to say, near the top centre of this photograph, which is printed with north at the top.

III. *Lunar Features*

(*top left*) The great craters Alphonsus (lower) and Arzachel (upper); the regular deep Alpetragius is seen to the right, with its massive central peak. Photograph by the late Dr. Dinsmore Alter, 60 in. reflector. (*top right*) The Mare Serenitatis, with part of the Apennine range. Linné is seen as a white patch, slightly below centre and to the left on the Mare. (*below*) The Messier twins, with the double 'comet' ray, on the Mare Fœcunditatis. Photograph by Commander H. R. Hatfield.

IV. *Craters on Mars, Mercury and the Moon*

(*top*) The surface of Mars, from Mariner 9. (*lower left*) Craters on Mercury, from Mariner 10
(*lower right*) The Mare Vaporum area of the Moon, photographed by Commander H. R
Hatfield on 20 May 1972 with a 12-in. reflector; the Hyginus Rill is seen, lower centre. The
comparison between Martian, Mercurian and lunar craters is very significant.

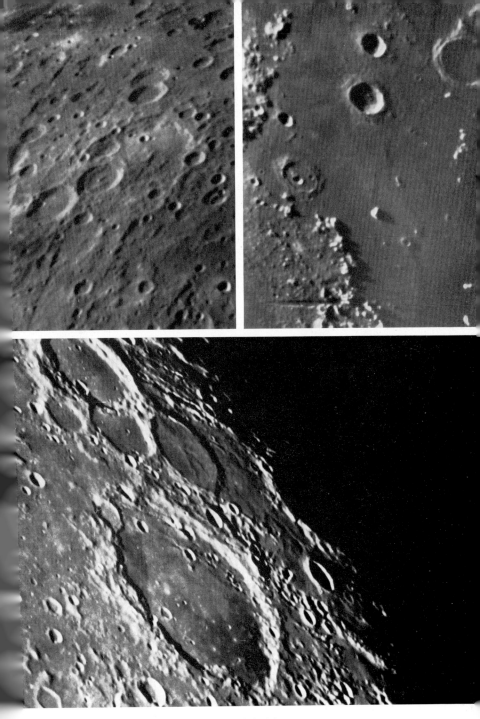

V. *Lunar Walled Plains*

(*top left*) The great ruined plain Janssen: H. R. Hatfield, 15 May, 1967 (12-in. reflector). (*top right*) Archimedes, Aristillus and Autolycus, with the Alpine Valley lower left; Hatfield, 17 April 1967, 12-in. reflector. (*lower*) Schickard and Phocylides, with the plateau Wargentin.

VI. *The Lunar Surface: Apollo and Luna*

(*top*) Apollo 12, showing the lunar module, Alan Bean, and the astronauts' footprints.
(*below*) Part of the lunar panorama transmitted by the Soviet automatic probe Luna 13 on
26 December 1966.

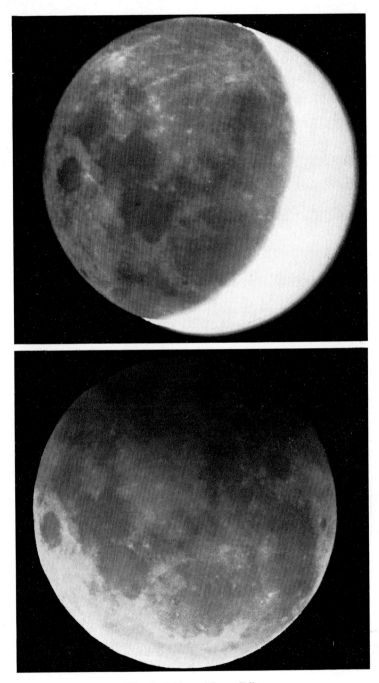

VII. *Earthshine and Lunar Eclipse*

(*top*) The Earthshine; P. W. Foley, 10 October 1974, 12-in. reflector. The sunlit area is necessarily over-exposed, and details on the earthlit portion are well shown. (*lower*) Total eclipse: 18 November 1975, 22.18 hours; Foley, 12-in. reflector. The Moon was only just fully immersed in the umbra, so that the lower (northern) limb remains relatively bright.

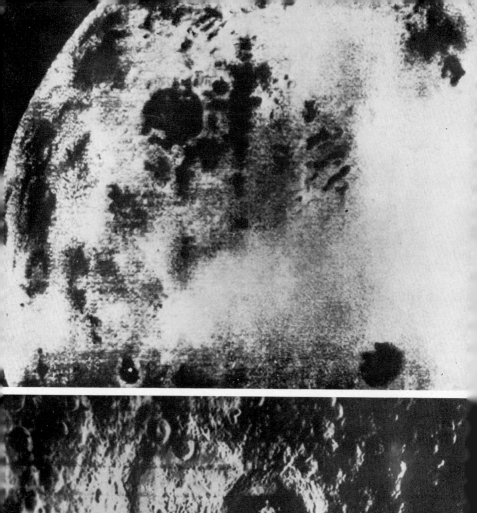

VIII. *The Far Side of the Moon*

(*top*) The Moon from Lunik 3, October 1959, showing part of the far side; Tsiolkovskii is seen at the bottom left. (*below*) Tsiolkovskii, as photographed from Orbiter 3; it intrudes into the larger, light-floored Fermi. The improvement in photographic technique is remarkable

IX. *The Moon from Mariner 10*

The N.E. part of the Moon; 3 November, 1973. This picture was taken by Mariner 10, *en route* to Mercury via Venus. The dark area near the centre is the Mare Humboldtianum, 125 miles in diameter. At the time, Mariner was 70,000 miles from the Moon.

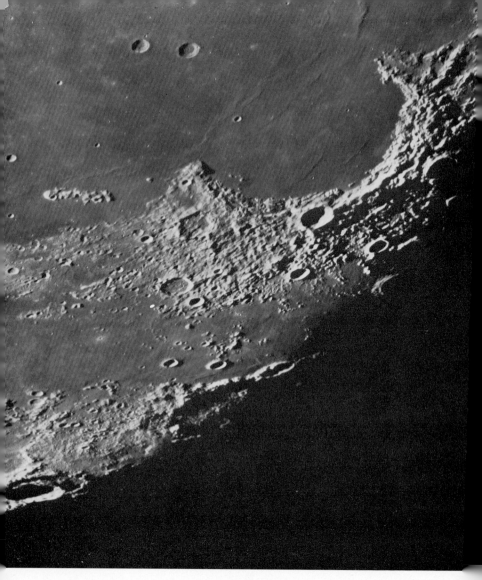

X. *The Sinus Iridum*

The Sinus Iridum; 23 December 1966, G. P. Kuiper. The Straight Range is seen to the centre left; the two craters to the upper left are Helicon and Le Verrier.

XI. *The Moon; almost full*

Photograph by P. W. Foley, 12-in. reflector, 6 March 1974.

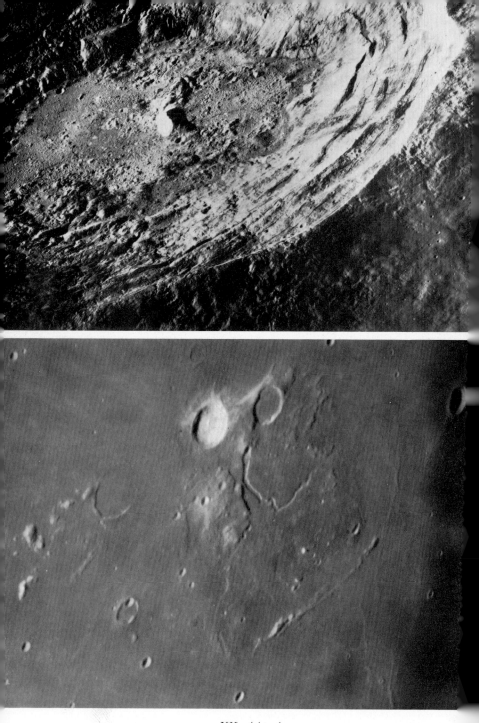

XII. *Aristarchus*

(*top*) Aristarchus, from Apollo 15. (*below*) Aristarchus as photographed from Earth; note i
brilliance compared with its neighbour Herodotus. The winding Schröter Valley is we
shown, and to the centre left is the incomplete crater Prinz.

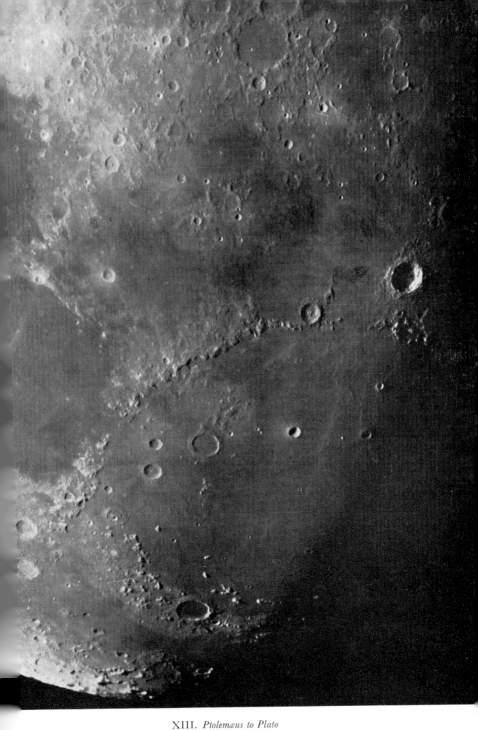

XIII. *Ptolemæus to Plato*

View of the area from Ptolemæus (*top centre*) to Plato (*lower centre*); Copernicus to the right, he Mare Serenitatis to the left. H. R. Hatfield, 27 January 1969, 12-in. reflector.

XIV. *The Moon from Apollo 16*

The Mare Crisium is shown to the upper left, Clavius to the lower left. Much of the area in the photograph is of the far side, inaccessible from Earth.

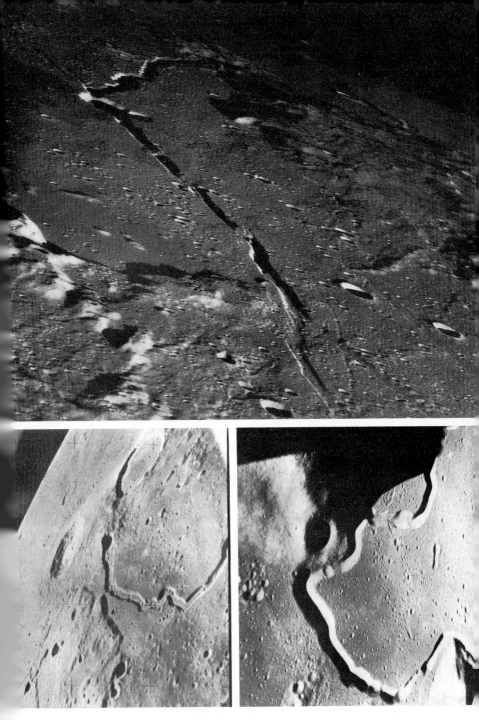

XV. *Lunar Rills*

top) Oblique view of part of the Hyginus Rill; Apollo (*lower left*) Aristarchus, Herodotus and Schröter's Valley from Orbiter. (*lower right*) Another Orbiter view of part of Schröter's Valley.

XVI. *The Moon; from Apollo*

(*top*) View from Apollo 14; the crescent Earth is seen above the lunar surface. (*below*) View
of the surface at the Apollo 17 landing site (Taurus-Littrow), showing the LRV and
Dr. Schmitt at the ALSEP.

Chapter Thirteen

EXPLORING THE MOON

THE APOLLO PROGRAMME was not merely a question of sending men to the Moon, in the same way as the first climbers aimed for the top of Everest. Scientifically it was immensely valuable; and without it all our ideas of going into space would have been set back for a very long time. But the human aspect was all-important, simply because machines are expendable while men are not. If an automatic probe crashes to destruction on the Moon (as many of them have done) the loss is annoying, but no more. Once crews are involved, the situation is entirely different.

The Russians have so far attempted no lunar flights, and their programme has been proceeding along different lines. Instead of sending men to the Moon, they have sent soft-landers, re-coverable probes, and 'crawlers'. To date, two there-and-back vehicles have been successfully sent to the Moon—Luna 16 in 1970 and Luna 20 in 1972; both came down in the eastern hemisphere, in the general area of the Mare Fœcunditatis and the highlands of Apollonius, and both brought back rock samples, though the quantities were naturally very limited as compared with the American hauls.

The crawlers, taken to the Moon in Luna probes, were in a different category. Lunokhod 1, looking distinctly like an antique taxi-cab, was landed in the western part of the Mare Imbrium in November 1970, and for months it moved around the surface, guided by its controllers in the U.S.S.R. and sending back valuable information; finally its power gave out, but no doubt it will be collected one day, because we know exactly where it is. Lunokhod 2 had as its target the semi-ruined crater-bay Le Monnier, on the eastern edge of the Mare Serenitatis; altogether it covered a traverse length of almost twenty miles. The region covered was only 110 miles north of the site of the Apollo 17 landing which had taken place a month earlier. Weird though they looked, the Lunokhods were brilliantly successful, and probably more of them will be dispatched in the reasonably near future.

Despite this, it is true that most of our detailed knowledge of the Moon has been American-gained, merely because nothing can take the place of an explorer actually on the spot. The astronauts were selected with the utmost care, and were given a training which was ultra-rigorous by any standards. They are not all alike (and I know most of them), but all possess more than the usual ration of courage and ability, and without exception they acquitted themselves impeccably. The crew of Apollo 14 included one of the original seven astronauts: Alan Shepard, who had made his first brief space-hop only ten years before. This, surely, shows how rapid the progress of space research has been.

With one exception the Moon-travellers were astronauts first and scientists afterwards, though they were given as much purely scientific training as was possible in the time allotted. The exception was Dr. Harrison (Jack) Schmitt, of Apollo 17, a professional geologist who had been trained as an astronaut so that his specialized knowledge could be used to the full. It is fair to say that most of the really important scientific information came from the last three Apollos, when the foundations had been well and truly laid.

The worst crisis came, of course, with Apollo 13. During the outward journey there was an explosion in the service module of the space-craft, which blew a gaping hole in the side of the vehicle and put the main engines permanently out of action. The lunar landing was promptly called off, and it was only by almost superhuman improvisation by the technicians at Mission Control, Houston, that the crew were able to come home safely. Yet had the astronauts been men of lesser calibre, they would have stood no chance at all, and the whole episode was a timely reminder that space research is extremely dangerous. Disaster would have been inevitable if the explosion had happened on the way back from the Moon, because the Lunar Module would have been jettisoned. In the event, it was the power in the Lunar Module which saved the situation.

I do not propose to say much about what is termed the hardware aspect, if only because I am not competent to write about it except in the most general terms. I will only repeat that the Apollo flights were essentially reconnaissance missions. They could do little more than take three men to the neighbourhood of the Moon, send two of them down to the surface for a limited period, and then come back. There was no provision for rescue in the event of a failure in the Lunar Module while it remained on the surface, and everything depended upon the successful, first-time functioning of that one ascent engine.

Quite apart from studies of the Moon itself, there were unrivalled opportunities for carrying out researches into other branches of astronomy. The lack of atmosphere, the absence of any appreciable magnetic field, and the low gravity make the Moon an ideal site. Consider, for instance, what we call the solar wind, which is made up of a plasma (that is to say, ionized gas) streaming out from the Sun in all directions. It hits the lunar surface, unchecked by any of the magnetic or atmospheric barriers which shield the Earth, and darkens the outer materials; it can be collected by means of suitable equipment, and samples of it have been brought back for analysis. It has even been said that 'the Moon collects the Sun for us'—because the solar wind is a sample of the Sun, and has the same chemical composition. Another interesting experiment was designed to detect gravitational waves coming from deep space, which may or may not exist (the evidence so far is very inconclusive). Of course the Apollo programme expanded as it went along, and it will, I think, be best to discuss it in roughly chronological order.

The main programme began just before the end of 1968, when Apollo 8 carried astronauts Lovell, Borman and Anders round the Moon, coming down to a mere 70 miles above the surface. Apollo 9 was an Earth-orbiting vehicle, in which the Lunar Module was tested at conveniently close range. Apollo 10, in May 1969, sent its Lunar Module to within ten miles of the Moon. (Colonel—now General—Thomas Stafford was commander on that occasion; he was also commander of the American team involved in the space link-up with a Russian vehicle in July 1975.) Apollo 11, in July 1969, was the first actual landing mission; Michael Collins remained in the command module of the vehicle, while Armstrong and Aldrin made the descent. There followed Apollo 12, later in 1969, and then, after the luckless 13, Numbers 14 to 17 inclusive. Apollo 14, with Alan Shepard as commander, came down in the Fra Mauro area; the explorers pulled their equipment along in a 'moon-cart', but with the last three missions the Lunar Roving

Vehicles or LRVs were used to great effect. The programme ended with Apollo 17 in December 1972. Eugene Cernan, the mission commander, remains—so far—the last man on the Moon, and there are no prospects of anyone else going there yet awhile.

I propose to deal only with actual lunar science, and to select only the main points, because I am writing for the lunar observer who will never be able to go to the Moon; but I cannot resist beginning with a quote from Colonel Edwin Aldrin, who came down the ladder from the grounded Module only a few minutes after Neil Armstrong's classic 'one small step'. Aldrin described the scene at Tranquillity base as 'magnificent desolation', which proved to be equally true for all the other landing sites.

By the time of Apollo, Gold's dust-theory had long since been disproved, and the Surveyor unmanned probes had sent back preliminary data concerning the surface materials. Volcanic rocks such as basalts were much in evidence, and this was fully confirmed by all the Apollo and Luna missions—not only *in situ*, of course, but from the many samples brought home for laboratory analysis. Not all the samples were alike. This was only to be expected, since the landing areas were deliberately selected so as to give as much variety as possible; for instance Apollo 15 came down in the foothills of the Apennines, while Apollo 16 landed in the bright uplands near Descartes.

Without delving at all deeply into geology I must, I feel, introduce a few technical terms. Basalts are fine-graded, dark igneous rocks composed chiefly of material called plagioclase (feldspar) and pyroxene, together with other minerals such as olivine; breccias are made up of fragments of rocks, minerals and glasses cemented together by the effects of heat and pressure, while the material holding them together is called the matrix. Magma is the molten rock below the surface, which becomes igneous rock when it solidifies and is termed lava when it comes out on to the actual surface. On the Moon, particles larger than about one centimetre in diameter are called rocks, while smaller particles are known as fines.

The outer surface layer of the Moon is termed the regolith. It is made up of fragmented material formed by the breaking-up of the underlying bedrock; it is this regolith material which makes up the matrix of the breccias. It is rather variable in depth—from four to five metres over the maria and about twice this in the highlands; in the Taurus-Littrow area visited by Apollo 17 it seemed to go down to over 30 metres in places, but in some areas in the Descartes region (Apollo 16) there was only a very thin layer. It is generally supposed to have been produced by the pulverizing action of meteoritic particles striking the Moon. There has been plenty of time for this to happen, since the Mare lavas solidified at a fairly early stage, and the upland areas are older still. There is definite evidence of 'churning', and some external action must be responsible.

A fascinating theory, again due to Dr. Thomas Gold, assumes that there was one tremendous flash of energy from the Sun which had a profound effect upon the surface materials of the Moon. According to Gold, this flash occurred between 30,000 and 100,000 years ago, and was due to an asteriod or a comet falling into the Sun; it swept away part of the Earth's outer air and virtually all the atmosphere of Mars, and glazed the surfaces of the Moon and Mercury. However, the theory has not met with much support, and the evidence in favour of it is decidedly slender. In any case, the regolith is quite firm enough to bear the weight of even a massive space-craft, and there now seems little fear of finding any dust-drifts deep enough and soft enough to be dangerous.

Like the Earth, the Moon has proved to have a crust, a mantle and a core (Fig. 41). One major surprise was that both crust and mantle are thicker than with the Earth, probably because the temperatures deep inside the Moon are much lower. The crust thickness is generally between 30 and 40 miles, though it may be more in places (remember that as yet we have positive information from only a relatively few sites). Below the crust lies the mantle, which goes down to 600 miles or so; below this again comes what is known as the asthenosphere, which may be a region of partial melting. The core of the Moon is thought to have a radius of something like 180 miles, and to be made up of iron; it may be liquid now, and in any case it must certainly have been so at some time in the past. The core temperature is probably about 1,500 degrees Centigrade, which is much less than that of the Earth, but is still appreciable; the old idea of a Moon which is cold throughout has been shown to

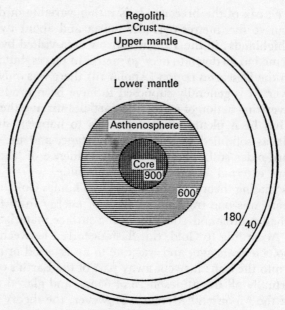

Fig. 41. A cross-section through the Moon (Figures
indicate miles below the surface)

be wrong. The later Apollos took equipment designed to
measure the amount of heat flowing out from inside the Moon.
The equipment taken by Apollo 16, the one highland site, was
accidently broken while being set up, and all attempts at
emergency repair proved abortive, but the Apollo 17 heat-flow
device worked well, and the results were quite conclusive.

If there is an iron core to the Moon, we might expect a
magnetic field, but in fact the overall field is negligible. On the
other hand, when the first rocks were brought back by Armstrong and Aldrin it was found—rather surprisingly—that the
lava samples possessed remnant magnetization, indicating that
between 3,000 and 4,000 million years ago the Moon had a
magnetic field which has since vanished. (This age can also be
expressed as 3 to 4 æons—one æon being equal to a thousand
million years.) The field could have arisen from a dynamo
process in the iron core, or else by magnetization of the deep
interior produced by a primeval Solar System magnetic field;
we simply do not know, and neither do we know just what

166

happened to it, though it is logical to assume that a small core will lose its magnetism more quickly than a large one.

Though the overall lunar magnetic field is so small, there are localized fields over large craters. These were studied from small sub-satellites sent out from Apollos 15 and 16 while the command modules were still in orbit round the Moon. The magnetization of the lunar crust is widespread, but for various technical reasons the lines of magnetic force escape only where there are large surface craters. The local fields seem to be strongest and most variable on the far side of the Moon, where the crust is appreciably thicker; the crater known as Van de Graaff is one such source, but there are plenty of others. In Van de Graaff the layer of magnetized material on the floor is probably about a kilometre thick, and is magnetized horizontally in an east-west direction. *En passant*, Mercury, whose surface looks very like that of the Moon and whose craters are presumably of the same type, has a distinct magnetic field, as was found by the probe Mariner 10 in 1974. Therefore, the two bodies are not so basically alike as might be thought from a casual glance; of course Mercury is rather larger (diameter 3,000 miles) and more massive, so that it may well have a more extensive iron core. Venus, which is about the same size as the Earth, seems to have a negligible magnetic field, and the same is probably true of Mars.

Our knowledge of the Moon's internal structure is drawn almost entirely from the seismometers or 'moonquake recorders' which have been set up at the Apollo landing sites. These work in just the same way as our own seismometers, but they can be made a thousand times more sensitive, because the Moon is so seismically quiet—we can, for instance, discount the effects due to passing lorries, or waves beating against the sea-shore! Seismic waves are of various kinds, some of which can travel through liquid while others cannot. This is how we have come to know the size of the Earth's liquid core with considerable accuracy. Natural tremors occur in the body of the Moon—I will have more to say about them in Chapter 15—but there have also been artificial shocks, and it is these which have told us that the so-called lithosphere, comprising the regolith, crust, upper mantle and lower mantle, is solid, rigid and stable down to a depth of more than 600 miles.

After Charles Conrad and Alan Bean, the Apollo 12 astronauts, had left the Moon and had returned to the orbiting part of their space-craft, the jettisoned Lunar Module was crash-landed back on the Moon; the impact occurred 47 miles from the point where the seismometer had been set up. Instead of recording a single jolt, the seismometer reported that there were reverberations which went on for some time, and the same thing was found with later missions—including Apollo 13, where the crash-landing of the last stage of the launching vehicle set up disturbances which continued for over four hours. It was even said that the Moon was 'ringing like a bell', though the analogy should not be taken too far. The effect was due to the reverberations of the seismic waves in the near-surface layers of the Moon, because of the low attenuation of the waves and the velocity structure in those regions. Undoubtedly the crust, below the regolith, is far from being perfectly uniform.

Much has been learned about the ages of the lunar rocks. One sample, from Apollo 17, seems to be about 4,500 million years (4·5 æons) old—comparable with the estimated age of the Moon itself; the great lava-flows began 3·9 æons ago* and ended about 3·1 æons ago, when the flooding of the Mare basins was completed. The dating of the rocks was achieved by radioactive methods. For instance rubidium, an element found in the rocks, decays steadily to end up as another element, strontium, and the relative percentages found together give a clue as to how long the process has been in operation. Obviously we can be positive only about the rocks which have been studied in our laboratories, and this is one reason why the landing sites for the Apollos were spread out so widely. There would be no sense in setting up two recording stations side by side; the spread has to be as great as is practicable, and the Russian Lunas have also helped.

Generally speaking, the average age of the highland crustal areas is now thought to be at least four and a half æons; the Mare basins were formed about four æons ago (Orientale is the youngest) and the lava-flooding followed. There is, of course, no doubt at all that the Moon's really violent period ended long

---

* Let me repeat—in astronomy, one æon is generally taken to be one thousand million years.

ago, and that any recent disturbances have been on a very mild scale indeed.

Among the features found in the lunar materials are vast numbers of small glassy particles, or 'beads', of microscopic size. Their colours are varied, and so are their forms. It is thought that they have been produced by what is termed shock metamorphism. A sudden, violent disturbance in the lunar material can cause alterations in structure, plus melting, and glass is the result. Meteorite impacts could certainly cause shock metamorphism, and so, probably, could some kinds of internal action.

The most startling of these beads were found by Dr. Schmitt during one of the expeditions from Apollo 17. I was in Mission Control, Houston, at the time; the date was 12 December, 1972. Schmitt and the mission commander, Eugene Cernan, had been driving in their Lunar Rover out to South Massif, one of the two mountains overlooking the valley, where the material was made up chiefly of breccias, with sub-floor rocks of basaltic nature. On the way back their route took them to a small crater which had been given the unofficial name of Shorty, and I can remember the throb of excitement when Schmitt's voice came through: 'It's orange—crazy!' There was a band of orange-coloured material circumferential to the crater, and Schmitt compared it with a fumarole effect, which would indicate relatively recent vulcanism. Samples were collected, but when they were brought home and analyzed the explanation proved to be very different. The orange colour was due to tiny coloured glassy beads, which were very old (about 3·8 æons), and fumarole activity was not involved. Yet it is strange that nothing comparable has so far been found anywhere else on the Moon—and the colour was so pronounced that it was easily visible on our television screens in Houston at the time when Schmitt was examining it, a quarter of a million miles away from us.

A few words about the Lunar Rover. This was a specially-designed vehicle, electronically powered, which was taken to the Moon inside the module and was then run out on to the surface. There were, in all, three Rovers used on the final missions; as I have said, Shepard and his colleague Mitchell, in Apollo 14, had had to drag their equipment along in a cart, while the explorers from the first two missions (11 and 12) had

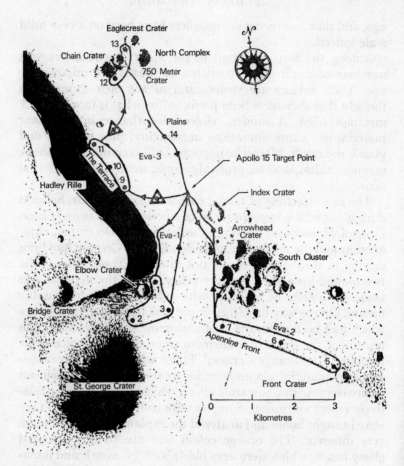

Fig. 42. Map showing the 3 EVAs (Extra-Vehicular Activity) of the
Apollo 15 astronauts

had to walk, which meant that they were unable to travel very
far. The introduction of the Rovers altered the whole situation.
For instance David Scott and James Irwin, from Apollo 15,
were able to drive right up to the rim of the tremendous Hadley
Rill (Fig. 42), in the Apennine area, almost three and a half
miles from their grounded module. The Rill is over 60 miles
long, almost a mile wide, and a third of a mile deep; there were
obvious signs of stratification, giving the impression that it
had been cut by a lava-stream in the Moon's active period.

The impressive peak of Hadley Delta was in full view, and proved to be quite unlike the rough, jagged mountains so often described by science-fiction writers in pre-Space Age days. Scott called it a 'featureless mountain'. The photographs make it seem much closer to the Rover than it really was; the distance to the base was over twenty miles. As all the astronauts have found, distances are difficult to estimate, partly because the lack of atmosphere means that there can be no softening of shadows, and partly because the Moon is smaller than the Earth, so that its surface curves more sharply and the horizon is correspondingly closer.

Surface coloration is generally very uniform. From Apollo 12, Conrad commented that 'the Moon is just sort of very light concrete colour. In fact, if I wanted to go and look at something I thought was the same colour as the Moon, I'd go out and look at my driveway.' This was in the Mare Nubium, close to the landing-point of the old automatic probe Surveyor 3. Indeed, Conrad and Bean walked over to the Surveyor and examined it; they found that 'it looked a light tan, as though dusty'.

It would take too long to give details of the various differences in composition of the rocks collected by the various Apollo missions, but one fact is plain: the rocks and fines are essentially volcanic. There have been major lava-flows, and meteoritic material is in short supply, though it is true that according to recent analyses the highland breccias contain evidence of ancient meteoritic débris. This paucity is rather surprising, because even if the maria and the principal craters are of internal origin there must be many small impact craters on the Moon— and to quote G. J. H. McCall, a leading expert on the subject, 'Where have all the meteorites gone?' This is a problem which remains to be answered.

No unknown substances were found on the Moon, and this too was predictable; but there are some minerals which are unfamiliar in structure, because the evolution of the lunar crust has taken place under unfamiliar circumstances. One such mineral has been named Armacolite, after the three astronauts of Apollo 11; *Arm*strong, *A*ldrin and *Col*lins.

This review of the findings of the Apollo programme is very sketchy and incomplete, as I well know. I have attempted to do no more than give the main highlights. Originally the programme had been scheduled to continue up to Apollo 21, but the premature ending was wise, partly because within its marked limitations Apollo had done almost everything of which it was capable, but mainly because if the journeys had continued there would have been a tragedy sooner or later. The failure of the single ascent engine in the Lunar Module would have condemned the explorers to death.

Much though we have learned, the Moon continues to surprise us, and in many ways the disagreements about its nature and its origin are as marked as ever. We should know much more by the time that the next astronauts set out. Meantime, let us turn to one of the most vexed questions of all—that of how the surface features came into being.

Chapter Fourteen

THE MOULDING OF THE SURFACE

SOME TIME AGO I was in my observatory, using my 15-inch reflector to show the half-moon to a group of boys who were interested in astronomy but did not pretend to know much about it. Their reactions were rather interesting. One or two boys said that the Moon looked like grey sand into which stones had been dropped; others compared it with porridge, but most said that the general view reminded them of dead volcanoes.

It is certainly true that the lunar scene is essentially volcanic in appearance, and a century ago few people doubted that the craters were of internal origin. Later, doubts crept in; there was a time, during the 1950s and early 1960s, when the volcanic theory was officially abandoned except by a few rebels (such as myself), and was replaced by the idea that the regular maria and the craters were formed by meteorites hitting the Moon. Today it seems that about 50 per cent. of lunar researchers support the impact theory, while the rest believe that the craters are essentially 'volcanic', using the term in a very broad sense. The whole problem has led to immense controversy, some of it strangely heated, and it has even been referred to as the Hundred Years' War. It is by no means over yet, even now that astronauts have been to the Moon. One cynic commented that 'in the Apollo results, everyone sees a confirmation of his own theories', and this is probably true.

I do not pretend to be unprejudiced. Impact craters must exist on the Moon, just as they do on the Earth; there should be a great many of them. So far as the maria and the large walled structures are concerned, however, I believe that the theory has so many weak links that we must look elsewhere. Before I give my reasons, let us pause to consider some of the less widely supported theories which have been put forward from time to time.

During the last century it was widely believed that the surface of the Moon was always extremely cold. (The only astronomer to give a more or less correct value was the fourth Earl of Rosse,

173

who lived at Birr Castle in Ireland; his results were not generally accepted until well after his death.*) This led to the Ice Theory, originally put forward by Ericson, of Norway, in 1885 and supported four years later by S. E. Peal, a tea-planter in Ceylon, who went so far as to write a booklet about it. Wrong though they proved to be, Ericson and Peal were perfectly sensible and logical—but this could hardly be said of the Austrian engineer Hans Hörbiger, who took up the idea in 1913 with the maximim possible energy. Hörbiger's book *Glazial-Kosmogonie* is one of the great classics of crank science, rivalled in modern times only by Immanuel Velikovsky's *Worlds in Collision* and its successors. It was Hörbiger who produced WEL (Welt Eis Lehre, or Cosmic Ice Theory) which became a powerful cult when Germany was ruled by the Nazis.

According to Hörbiger, everything apart from the Earth is made up of ice. The Moon's glacial covering is 150 miles deep, and the same is true of the planets (the canals of Mars are simply cracks in the upper ice-sheet). As the planets move round the Sun they are braked by the low-density hydrogen medium, and the same is true of the Moon, which is spiralling slowly downward and will hit us in the foreseeable future—long before the Earth itself plunges into the Sun and is snuffed out, producing a large sunspot. Also, the Moon is not the first of our satellites. There have been at least six previous moons, all of which have eventually collided with us, causing cataclysms which may be linked with past events such as the abrupt disappearance of the dinosaurs, seventy million years ago. Our present Moon was captured by the Earth between 13,000 and 14,000 years ago, and led to earthquakes and eruptions, one of which submerged the island of Atlantis. Another episode linked with the theory is, needless to say, the Biblical Flood.

How anyone could take this rigmarole seriously is hard to understand, but plenty of people did so, and during the 1930s the German Government had to issue a statement to the effect that it was still possible to be a good Nazi without believing in

---

* There were two famous Birr astronomers; the third Earl of Rosse, who built a 72-inch reflector in 1845 and used it to discover the spiral nature of the galaxies, and his son, who measured the heat of the Moon. The story is an incredible one; I have described it elsewhere—in *The Astronomy of Birr Castle* (Mitchell Beazley, 1973).

WEL. Moreover, there was one very serious lunar observer who became a Hörbiger devotee. This was Philipp Fauth, who published an elaborate map of the Moon and wrote several books on the subject. Fauth, like Peal so long before, believed the craters to be lakes of frozen water, and he agreed with Peal's words: 'As the lakes slowly solidified in the cooling crust, the water vapour rising from them formed a local, dome-shaped atmosphere, which became a vast condensed snowy margin and piled as a vast ring.' The maria, then, were actual sea-surfaces which had solidified.

Fauth, unlike Peal, thought that the ice had come from space in a sort of cosmic storm. He discounted the measurements which show that at noon the temperature at the lunar equator can exceed 200 degrees Fahrenheit, which does not seem very suitable for the permanent existence of either ice or snow; he also disregarded the obvious fact that an icy crater-rampart would not keep its shape for long, and would flatten out under its own weight. Fauth died in 1943, and so far as serious scientists were concerned WEL died with him. By now we can even discount the idea of sub-crustal ice. As recently as 1961 Z. Kopal of Manchester suggested that there could be glaciers at shallow depth, and that the famous lunar domes could be nothing more than glaciers covered with dust, so that the central pits of the domes represent the springs of still-active geysers, but the theory met with scant support, and Apollo has shown that it is completely wrong.

The ice theory may seem peculiar, but an even more remarkable idea was proposed in 1942 by a certain Herr Weisberger, or Vienna, who solved the whole problem very neatly by denying that there are any mountains or craters at all. He attributed the surface markings to storms and cyclones in a dense lunar atmosphere, and was most offended when the astronomical world failed to treat him with due respect.

It was of course difficult to surpass Herr Weisberger, but a Spanish engineer, Sixto Ocampo, did so in 1949, when he announced the Atomic Bomb theory. After explaining that the Moon does not rotate on its axis (!), Señor Ocampo went on to prove that the Moon used to be inhabited by technologically-advanced beings who indulged in an orgy of nuclear war, and destroyed their civilization, producing huge craters in the process. He added that the craters were of different types because the opposing sides used different types of weapons, one species of bomb producing a crater with a central peak and another kind a crater with a flat floor. The Alpine Valley and the Straight Wall were engineering works. After the fall of the bomb which produced the ray-crater Tycho, the lunar seas were 'fired' and expelled at great speed, falling back on the Earth and causing Noah's flood.

Señor Ocampo presented his paper to the Academy of Arts and Sciences at Barcelona, and was most annoyed when they declined to publish it. It was eventually printed in a small South American periodical, together with a letter in which the author complained that an unscrupulous British writer had stolen his theory and was planning to publish it as original work, thereby depriving Spain of the glory of the discovery. Ocampo's life-work was thereby brought to a successful conclusion, and he died almost immediately afterwards.

It should not be thought that Herr Weisberger and Señor Ocampo have had the monopoly of weird ideas. Remember, there is still an International Flat Earth Society whose members maintain that both the Earth and the Moon are shaped like gramophone records, while the members of the German Society for Geocosmical Research believe that the Earth is the interior of a hollow globe, so that the Sun is inside it. (I once put these two societies in touch. The resulting correspondence was most enlightening, but hardly relevant here.) Space forbids any discussion of the theories of D. P. Beard, who in 1917 proposed that the craters are coral atolls, or Dr. Immanuel Velikovsky, still something of a cult-figure in America, who believes the planet Venus to be an ancient comet which bounced about the Solar System in past ages, periodically by-passing the Earth and Moon, and causing phenomena such as (of course) the Biblical Flood. Even professional scientists are not immune; M. W. Ovenden, former Secretary of the Royal Astronomical Society, wrote a book some years ago in which he claimed that there can be no large lunar-type volcanic craters on Earth, as they would soon be rubbed away by friction as the Earth rotates beneath its atmosphere. The list is almost endless, but it is perhaps time to turn to ideas which, even if certainly wrong, do at least have something of a logical basis.

There are, for instance, various tidal theories, of which the best known was outlined by the Bulgarian astronomer, N. Boneff in 1936. According to Boneff, the craters were formed when the Moon's crust had just solidified, and the Moon, much closer to the Earth than it is now, was rotating on its axis comparatively quickly. The hot, viscid interior was much more affected by the tidal pull than the thin crust, and so at each rotation of the Moon on its axis the molten lava surged upward, breaking through the weak points in the crust. The action was rather like that of a pump, so that gradually the large craters were built up. As the Moon receded and the axial spin slowed down, the tidal effects lessened, and the formations produced were smaller. At last the crust became too solid to be broken by the surging magma inside, so that crater-building ceased altogether.

Boneff explained that the Earth's crust was not then solid enough to register any similar craters, though he did not rule out the possibility that the Moon still influences the numbers of earthquake shocks. Moreover, if it is agreed that the Moon will one day approach the Earth once more, Boneff maintains that it may yet be capable of covering our lands and drying sea-beds with craters before it is itself torn apart by the Earth's pull. The last paragraph of his paper reads: 'An Earth without a Moon, surrounded by a ring of minute bodies and entirely covered with formations of the lunar type, except perhaps at the poles—that is the probable state of the Earth-Moon system, if it still exists, after many thousands of millions of years.'

Luckily, this sombre picture need not be taken very seriously. Apart altogether from the fact that the time-scale is wrong, the whole theory is unsound. The Earth's 'danger-zone' was also mentioned by Kopal in a theory published in 1966. This time the Moon is said to have approached the Earth about 1,800 million years ago, so that the Earth-facing hemisphere came within the Roche limit while the 'back' hemisphere did not, and the large maria on the familiar hemisphere were produced by mechanical damage on the affected side only. Yet it seems that the effects all over the Moon would be noticeable, and there would be no sharp cut-off, so that this idea too has met with no support.

In 1929 Ingolf Ruud, of Norway, proposed the 'direct contraction' theory, according to which the crust of the Moon shrank around a less-yielding interior and thinned and stretched at its weakest points, with the formation of circular craters. The smallest formations would be the oldest, and the greatest of all the circular plains, the Mare Imbrium, would be the youngest. There are fatal objections here, as there are also in the idea put forward later by the French mathematician A. Fillias, in which the Moon's interior was said to expand against a firm crust. Finally, I must make brief mention of an idea by K. H. Engel, a Czech astronomer who spent most of his life in America, and who maintained that the craters and other features were formed by the spontaneous solidification and crystallization of a fairly shallow lava-layer.

Having 'cleared the air', so to speak, there remain the two main theories of today, the impact and the volcanic—better referred to as the exogenic and the endogenic. Since I do not pretend to be impartial, it is only right to give first innings to the meteoritic hypothesis, according to which the craters (and, for that matter, the main seas) were produced by the cosmical bombardment of the Moon in past ages.

The idea was originally put forward in 1824 by Franz von Paula Gruithuisen, of 'dark gigantic ramparts' fame. It was then more or less forgotten; revived briefly later in the century by R. A. Proctor, a great English popularizer of astronomy who was also a skilful researcher; and then discarded again, even by Proctor himself. Next, in 1892, came a famous paper by G. K. Gilbert, one of the leading geologists of the time, in which the theory was treated in more detail. But the modern revival is due largely to the work of Dr. Ralph B. Baldwin in the United States, who has written two major books on the subject as well as many papers and other contributions. From what follows here, it is obvious enough that I disagree with much of what Baldwin has proposed, but this is not to suggest that I have anything but the greatest admiration for him and his work. His reasoning is precise and clear-cut; his books and papers are presented in impeccable fashion, and he has made numerous contributions to lunar study which are of fundamental importance. It may well be that in the end he will be proved right and people such as myself will be proved wrong. Meanwhile, we agree to differ; and there can be no harm in this, as is, I feel, shown by the many cordial discussions which we have had over the years. There cannot be the slightest doubt that Baldwin will always be remembered as one of the greatest pioneers in lunar research.

Following the publication of Baldwin's first book, in 1949, the impact theory became fashionable—as indeed it still is, particularly in America. Those who came out strongly in favour of it included H. C. Urey, T. Gold, E. J. Öpik and the late G. P. Kuiper, though all these authorities disagreed over matters of detail. Urey, in 1956, wrote that it was no longer necessary to go on discussing the merits of impact *versus* vulcanism, since the whole question had been settled. Yet with the greatest respect to a Nobel Prize winner and one of the world's leading scientists, I suggest that matters are not quite so clear-cut as this. Even the most ardent supporter of the impact hypothesis must admit that there are horribly weak links in the chain of argument, and it is agreed that even if the regular maria were produced by falling meteorites they were certainly filled with lava at a later stage. There is thus no escape from the conclusion that the Moon was once the scene of tremendous volcanic activity.

One alleged objection to the impact theory, however, turns out to be no objection at all. It has been claimed that a meteorite would cause a circular crater only if it fell straight down, and that a missile landing at an angle would produce an elliptical scar. This is wrong, as practical experiments prove. On arrival, the meteorite would penetrate the Moon's crust, and would act in the manner of a powerful explosive, so that a circular crater would be formed. Neither is there anything wrong with the time-scale. Even the youngest of the major craters are probably a thousand million years old (that is to say, one æon) and most of the rest are older still.

Also, there are impact craters on Earth, and much can be learned from studies of them. The best-known example is the Arizona Crater, sometimes called the Barringer and sometimes the Coon Butte, which is visible from Highway 66, the main road running between Flagstaff and the town of Winslow. It has a diameter of 4,150 feet, with a maximum depth of 700 feet; the wall rises to 150 feet above the surrounding desert. It is an impressive structure, and is well worth visiting. It does not take long to walk down to the bottom of the floor—though the return climb, under a broiling Arizonan sun, can be somewhat tiring, as I have found!

There is no doubt that the crater was due to the fall of a meteorite (though, piquantly, Gilbert regarded it as 'a steam explosion of volcanic origin'). Many meteorite fragments have been found nearby, and one estimate gives the mass of the original missile as over a million tons, with a probable diameter of 200 feet. The date of the event is not known, but it was certainly prehistoric, and took place more than 10,000 years ago. Another crater which is almost certainly meteoritic in origin is Wolf Creek, in Australia. But when we consider other structures which are claimed to be due to impact, doubts begin to creep in.

For instance, there are some structures in the Canadian Shield, of which the largest is known either as the Chubb Crater or the New Quebec Crater. All these have been widely attributed to impact, but they seem to be related to the general tectonic patterns in the area, which may be highly significant. The Chubb Crater was found in 1950 by the Canadian prospector after whom it is named, and was described by him as being 'an immense hole looking like a great teacup tilted at an angle'. It is two miles across and 1,500 feet deep, while part of the floor is occupied by a lake. Its age is very uncertain, but it must be many thousands of years old, whatever may be its origin.

Then there is the Vredefort Ring, in South Africa, not very far from Pretoria. Here we have a well-marked structure, inside which lie the towns of Parys and Vredefort itself. Driving into the Ring, one is very conscious of the circular hills which make up the wall, and when I flew over it in a helicopter recently I saw that the general form of the crater was very obvious indeed. There are also some small features known as shatter-cones, which are produced by sudden, violent shocks in the ground, and which were once—wrongly—believed to be peculiar to impact craters. Geologists who have made a close study of Vredefort are unanimous in saying that it is of internal origin. There are many reasons for this opinion, and all point to the same inevitable conclusion. For one thing, the Ring lies exactly on top of a block of volcanic rock which is quite definitely older, and to imagine that a missile could land obligingly just in this position is stretching coincidence too far. It would have to be a very discriminating meteorite.

In fact, it may well be that many of the alleged impact craters in the official lists are not meteoritic at all. In any case, impact structures on Earth are small and rare; even allowing for the destructive effects of erosion by wind and water, one might reasonably expect to find more of them. Neither do we know of any which have been produced in historical times, and the only major meteoritic falls of our own century, those of 1909 and 1947, failed to produce anything even remotely comparable with Barringer or Wolf Creek. The Siberian missile of 1909 may well have been the nucleus of a small comet, which would have been icy in nature and would have evaporated; the force of the impact blew pine-trees flat over a wide area of the Tunguska region (fortunately without causing any loss of life, for the excellent reason that nobody lived there), while the 1947 meteorite, which came down in the region of Vladivostok, broke up during its headlong descent through the atmosphere and landed in fragments, producing nothing more startling than a number of pits. It is unfortunate that we have no large 'modern' impact craters to provide us with reliable information. Those produced by artificial explosions are not really significant, because of the vast difference in scale.

Even the proved terrestrial impact craters are puny by the standards of the Moon, and neither are they truly 'lunar' in form. Lunar craters are shallow relative to their diameters, and are more like saucers than mine-shafts. For instance, Theophilus has a diameter of 64 miles, and an average wall-height of around 2·7 miles. 64 divided by 2·7 is about 24, which is the depth-diameter ratio; for Ptolemæus the figure is 112, and for the 12-mile Bessel, on the Mare Serenitatis, it is 15. The depth-diameter ratio for the Barringer Crater in Arizona is rather less than 6, though I agree that this takes no account of the effects of erosion over the ages.

In his original books, Baldwin placed great stress on the fact that when the depth-diameter ratios are plotted for lunar craters, meteorite craters and bomb craters, and set out in a graph, the result is a smooth curve; the smaller the crater, the greater the depth relative to the diameter. This is quite true, but as terrestrial volcanic craters fit equally well into the same pattern it provides no evidence either way.

One vitally important question concerns the nature of the seas. The maria are of two main kinds: regular (Imbrium, Serenitatis and Crisium, for example) and irregular (Nubium, Frigoris, Australe). Long ago N. S. Shaler suggested that the maria were due to impact and the craters to vulcanism; this means introducing two processes where one will do, but we have to decide whether the maria are essentially similar to the walled plains or whether they come into an entirely different category. Let us first see how the impact theory supporters explain the sequence of events.

Begin by assuming that in its early period the Moon developed a comparatively firm crust, from which relatively little of the inner magma could flow out. There was not much cosmical bombardment, so that the situation was quiescent. Then, rather more than four aeons ago, the bombardment started. Huge basins were formed; Serenitatis, Crisium, Nectaris and Humorum were among the first, with ages of between 4·1 and 4·2 aeons. Then, about 3·9 aeons ago, came the greatest impact of all, producing the vast Imbrium basin and having profound effects all over the Earth-turned hemisphere of the Moon. The crust was ruptured time and time again, and magma began to seep out. There was one more huge impact about 3·8 aeons ago, resulting in the complex ringed structure of the Mare Orientale, and then no more; but tremendous lava-flows had already started to pour forth, filling the Mare basins and producing the seas which we know today. The Moon's rotation had already become synchronous when the main bombardment started, and of course there were impacts on the far side as well, but there was less outpouring of lava, because the crust on the far side was thicker and less easy to rupture, leaving the basins empty and light-floored. The lava-flow period lasted from 3·9 to 3·1 aeons ago, and then stopped. Meanwhile, smaller meteoritic falls had produced craters and walled plains, some of which were filled with Mare material (as with Grimaldi and Plato), while others —presumably of later origin—were not. As the outpouring of lava slackened the craters were left relatively undamaged, and the youngest of them, such as Tycho and Copernicus, are unfilled, with their central mountain structures clear-cut. Ray-craters are naturally the most youthful of all the major structures, as is shown by the fact that the rays cross all other formations. Small volcanic craters such as those forming the chains of Hyginus and Rheita were produced at a relatively late stage.

It has also been suggested that there were some basins formed even before the cataclysmic events of about 4·2 æons ago, and that these also were filled during the lava-flow period, thus explaining the maria such as Frigoris and Australe, which are of no particular shape.

It all sounds delightfully plausible, but it would surely lead to a distribution of the maria, walled plains and craters which would be essentially random, and this is not what we find. So let me next turn to some objections to the impact hypothesis—with the full knowledge that I am courting criticism.

Any casual glance at a lunar map (Fig. 43) will show that the major formations tend to be arranged in chains or groups. There is a chain down the central meridian of the Earth-turned hemisphere made up of the Walter group, the Ptolemæus group, and perhaps Clavius in the south and Archimedes and Plato in the north—we might even include the Mare Imbrium, though I agree that this is questionable. There is a chain down the east limb, extending from Vlacq down to Endymion, and including the Mare Crisium; another down the west limb, reaching from Bailly to Schickard, Grimaldi and Pythagoras. These are shown on the chart, which is a perfectly fair one, since the craters shown on it have been selected only by size. (A Mercator chart is even more informative, though the inevitable distortions make it less convenient for this sort of work.) No such chains extend from east to west, and it seems, then, that we are dealing with formations which were born along lines of weakness in the Moon's crust, connected with the influence of the Earth; the lining-up of the chains with the central meridian can hardly be due to chance.

It may be argued that the craters really were produced by chains of meteorites which fell at the same time. However, Baldwin himself has pointed out that the craters in any particular chain are of widely differing ages—for instance Vendelinus in the eastern chain is clearly older than its neighbour Langrenus—so that this is really no loophole. In one comment, Baldwin maintained that the chains themselves were due to

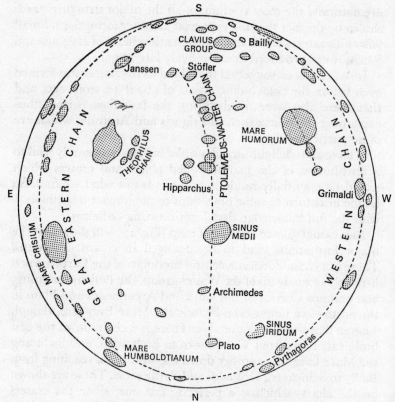

Fig. 43. Distribution of large lunar craters

tricks of the eye. When the Sun rises over a formation, the floor of the crater is thrown into shadow, and the crater itself is at its most conspicuous; the Ptolemæus group, for instance, is striking at the time of half-moon. This, according to Baldwin, led to a false impression, since chains aligned in an east-west direction would not catch the eye nearly so strongly, but in my view the chart given here speaks for itself.

Things are rather different on the Moon's far side, but here too we have evidence of chaining; note the great curve of large walled formations which runs from Birkoff through D'Alembert, Campbell, the Mare Moscoviense, Mendeleev, Gagarin, Mare Ingenii, Leibnitz and Apollo. And the immensely complex, ringed Mare Orientale, Hertzsprung and Korolev may well be

184

associated; there is nothing quite like them anywhere else on the Moon.

Note that I have included the regular maria in my analysis. It does seem definite that they are associated with the distribution patterns of the craters, which indicates—to me at least—that their origins are the same; and if we can show that the craters are endogenic (that is to say, produced by internal action rather than bombardment), then the regular maria must be endogenic too. This does not rule out the idea that they were filled with lava well after their actual formation as basins, and no doubt the Imbrian event had a tremendous effect here, since it caused the greatest upheaval ever known on the Moon. Z. Kopal has suggested that even among the craters there are fundamental differences in origin, so that some (such as Tycho) are due to impact while others (such as Alphonsus) are not; but in this case there would surely be a clear difference between the two types, and this does not fit the facts.

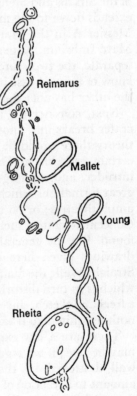

It is not only the large craters which arrange themselves in chains. There are strings of very small craters in many parts of the Moon, and these can only be of internal origin. Consider, for instance, the Rheita Valley (Fig. 44). Urey, in 1952 and 1955, maintained that it had been produced by a meteorite slicing through the rocks, but since it is a well-marked crater-chain, and not a true valley at all, the explanation is inadmissible. There are many cases of structures which are part crater-chain and part cleft; the Hyginus Rill is one. Moreover, there is a steady gradation from the 'strings of beads' up to formations such as Vogel, in the Hipparchus area, which consists of several craters which have run together.

Even if we suppose that we are dealing with strings of meteorites falling

Fig. 44. The Rheita Valley

obligingly in a line, the result would still be wrong. There would be a generally disturbed area, certainly not a chain of confluent craters of which some retain their separate walls. But there is no point in going into further detail, since even the most fervent impact-supporters agree that the crater-chains are not due to meteoritic falls. The Alpine Valley is in a different class; but here we also have to reckon with the parallel and transverse valleys, which from Earth are not easy to see, but which definitely exist. I first drew attention to them as long ago as 1949, and they are clearly shown on the Orbiter and Apollo pictures.

Groups and pairs of craters are very common. There are separate 'twins' such as Aliacensis and Werner, and joined ones of the Sirsalis and Steinheil types. The same sort of arrangement extends down to the much smaller craters, such as Messier and Messier A in the Mare Fœcunditatis and Beer-Feuillé in the Mare Imbrium. Note also that with the twins, either joined or separate, the two craters are always of the same kind. I do not know of a single case in which one twin has a central peak and the other has not; the two are always basically alike.

Next, consider the almost innumerable cases in which one crater breaks into another—and this is crucial; on the impact theory there would be randomness, on my endogenic theory a certain orderliness. The rule is that the smaller crater is the intruder, and the larger one the sufferer. Of course there are great formations which have much smaller craters inside their amphitheatres; or on their walls; Clavius is a classic example. Concentric craters such as Vitello and Taruntius are also to be found. But the general principle is best illustrated by the two drawings given here (Fig. 45). One is of Thebit, near the Straight Wall; the main crater is broken by a second (Thebit A) which is in turn disturbed by a third (Thebit D). The second is a freehand sketch, since it could not be drawn from observation; nothing quite like it exists on the Moon.

There are a few exceptions to the rule. The Sinus Iridum may be one; if we regard it as a crater, it has had its seaward wall destroyed by the Mare Imbrium. But the exceptions amount to a fraction of one per cent., and here again the impact theory fails. According to it, we would have a great many departures from the rule, since it is impossible to believe that all

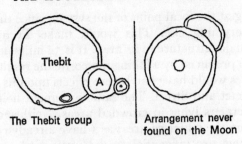

The Thebit group | Arrangement never found on the Moon

Fig. 45. *Crater arrangement*
(*left*) Thebit: small craters intruding into larger; (*right*) A large crater intruding into a smaller peaked crater in a way which is unknown on the Moon

the large meteorites fell first. On the volcanic principle there is no objection at all; the earliest outbreaks would be the most violent, and smaller craters would be produced later, when the Moon had calmed down.

There is, moreover, another point. I first discussed it in 1960, but at the time I did not realize that it had not been appreciated; subsequently Kopal mentioned it independently. To explain it, let us look again at Thebit.

Suppose that Thebit itself were formed by a meteorite impact. Later, a second meteorite fell on or very near the wall, and produced the crater A. If so, there would have been a major disturbance in the lunar crust, and the wall of Thebit would have been distorted for miles around the point of the second impact. This is not what is seen. Thebit's wall remains perfect and undamaged right up to the point of junction, and with other formations it is always the same; never is there any sign of ruin or débris. Could an impact-produced Theophilus have failed to shake down long stretches of the wall of the older Cyrillus, and could Tralles have been similarly merciful to Cleomedes? I doubt it; and this indicates that the craters were produced neither by explosive impact nor by explosive vulcanism. The process must have been much gentler.

Next, there are the rays which come from Tycho and other craters. These are surface features, as we know from the astronauts who have been to ray-affected areas. On the impact theory, the rays must have been formed at the same time as the

187

crater lying at the focal point of the system, since the streamers overlie everything else. This would make Tycho the very youngest major structure in its area. It is 54 miles across, and if it had been produced by a falling meteorite the resulting ground disturbances would have had severe effects upon its neighbours such as Pictet and Street. The same argument holds good for other ray-craters lying in crowded areas. Neither do the rays diverge from the crater-centres; as I have already commented, some of them are tangential to the walls, and round Tycho there is a ray-free zone.

Also, it is hard to see how the material could have been spread out in such long, straight lines. One of the Tycho rays, crossing the Mare Serenitatis, is often said to be 'reinforced' at Bessel, but whether the two rays are really one and the same is a matter for debate.

I have already mentioned the domes, which, like some of the rounded hills, are crowned by small, symmetrical pits. These were attributed to sheer chance hits. In 1949, when about a dozen such objects were known, Baldwin wrote that there should be just about this number: 'The agreement is excellent. The case for the meteoritic nature of the lunar craters is even more solidly founded, and the volcanic hypothesis is correspondingly weakened.' This did not satisfy me at all. I was furtunate enough to be able to use one of the world's largest refractors—the Meudon 33-inch, at the Paris Observatory— and after a few weeks' work I had found more than thirty summit pits. So many more have been revealed by the Orbiter and Apollo photographs that chance strikes can definitely be ruled out.

Both summit pits and central peaks present obvious difficulties to any impact theory, but would be expected if the craters are of internal origin. Here, of course, we have many terrestrial examples. One, which I know well, is Hverfjall, near the village of Reykjahlid in the Lake Myvátn region of Iceland. It is visible from the main road, and is quite a landmark. It is circular, with walls which rise to between 200 and 300 feet above the floor; it is three-quarters of a mile across, and has a central peak 80 feet high. The inner and outer slopes of the wall are quite gentle, and I found that I could more or less walk up them; no true climbing was needed. Close by are Lúdent, which has an intruding crater, and the smaller Hraunbunga. There is absolutely no doubt about their volcanic origin, and there is at least a superficial resemblance to the craters of the Moon, although the scale is different. I doubt whether many people would deny that Hverfjall is much more 'lunar' than the Barringer Crater in Arizona.

This, then, is a summary of my case against the impact theory for the major maria, walled plains and craters, though I must again stress that I do not for one moment doubt the existence of many small meteoritic craters on the Moon. I may well be wrong. Meanwhile, let us look at some possible alternatives to the whole idea of cosmical bombardment.

The first serious attempt at a comprehensive volcano theory was made in 1874 by two English amateurs, James Nasmyth and James Carpenter. (Nasmyth's telescope is now on exhibition at the Science Museum in South Kensington. He was also the inventor of the steam-hammer!) In their book, which has become something of a classic, Nasmyth and Carpenter pictured a central volcano erupting violently and showering débris round it on all sides, so that the matter ejected from the central orifice built up the crater wall (Fig. 46). As the eruptions became less violent, inner terraces were formed, and in the

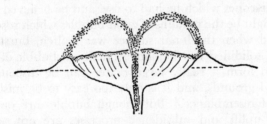

Fig. 46. Nasmyth and Carpenter's Volcanic
Fountain

dying stages of activity, when the explosions were only just powerful enough to lift the material out of its vent, the central peak was built up. Craters without central peaks were produced when the explosions ceased suddenly, so that the floor was covered with lava which welled up from inside the Moon.

The idea looks intriguing at first sight. The terraces, the

hill-top craters, the flooded plains and even the famous plateau Wargentin are accounted for, and the ringed formations do give the superficial impression of having been built up in this way. Unfortunately, there are any number of fatal objections. It is beyond all belief that a circular wall over 100 miles in diameter, and sharply defined (as with Clavius, for instance), could have been formed in this way; the slope-angles are wrong, and the walls are far too massive. Moreover, in a lunar crater the central peak is always considerably lower than the rampart, which would not be expected on the fiery fountain theory. The explanation given for the bright rays is equally untenable. Nasmyth and Carpenter thought that the crust of the Moon had cracked in places, much as a glass globe does when it is struck, and that lava had oozed out of the cracks, forming the rays. This may have sounded convincing in 1874, but it certainly does not in 1976! Yet although Nasmyth and Carpenter were so completely wrong, they rendered lunar science a service by publishing their book at a time when interest in the Moon was at a comparatively low ebb.

Much earlier—in 1665—an entirely different process had been suggested by no less a person than Robert Hooke, contemporary and rival of Newton. Hooke made some drawings of lunar craters which were remarkably good considering the low-power telescopes which he had to use, and he believed that the craters might be the remains of great bubbles which rose during the period when the lunar surface was molten, bursting and leaving a solidified rim behind. Actually, a bubble of the size needed to form a really large crater can be discounted on dynamical grounds, and it is only too easy to be misled by a superficial resemblance.* But though bubbles are out of the question, uplift and subsidence processes are not, and this brings me on to the caldera analogy.

Up to now I have been using the term 'volcanic' rather loosely; what I really mean is 'produced by internal action', i.e. endogenic. There are many small features on the Moon which look like true volcanoes, but the main formations are

---

* At a very serious NASA conference in the United States, I was once guilty of showing a slide of a 'lunar scene' which was, in fact, a photograph of the top of a layer of boiling coffee. I understand that at least some members of my audience were momentarily puzzled!

utterly unlike our Vesuvius or Etna (or, for that matter, structures such as the Olympus Mons on Mars), and they may well be more in the nature of calderas. A caldera is a volcanic crater formed by the collapse of the surface into an underground cavity, either by the rapid eruption of large quantities of magma or else by the withdrawal of magma from a magma-chamber. There is a caldera on the top of Vesuvius, known as Monte Somma, and many much larger ones in Africa, associated mainly with the Great Rift Valley in Kenya.

The idea that the lunar craters (and the regular maria) are nothing more nor less than large caldera-type structures was developed during the years following the last war by J. E. Spurr. It has been extended more recently, notably by Dr. Jack Green in America and Professor S. Miyamoto in Japan, and has a great deal to recommend it.

Large calderas may have arisen at a very early stage, producing basins which have now been obliterated in their original form even though the depressions were later covered by lava-flows to produce the irregular seas such as the Oceanus Procellarum. Then came the formation of the present-day regular maria, beginning with Mare Nectaris and similar basins and ending with Imbrium and, finally, Orientale. Crater-building began by the same kind of process, and this time we have no difficulty in explaining the chains and the alignment of the major structures with the central meridian—because the Moon's rotation was already synchronous, and lines of weakness in the lunar crust were influenced by the gravitational pull of the Earth.

Many of the early walled plains were simply miniature maria; let me repeat that Grimaldi, for instance, would have been ranked as a small sea if it had been better placed for observation from Earth. With these lava-flooded craters, any central peaks would have been destroyed by the welling-up of the lava, though later craters, formed when the crust had become thicker, retained central peaks and unflooded interiors. There is no trouble in explaining the chains, the occasional plateaux of which Wargentin is the best example, or the twins and groups; we can also account for the bays such as Hippalus and Doppelmayer on the Mare Humorum and Le Monnier on the edge of the Mare Serenitatis, where the seaward walls have been much reduced in height and are now discontinuous. The mountain border between the Mare Imbrium and the rather older Mare Serenitatis had also come in for much rough treatment, but was so massive that it had managed to survive except for one stretch between the modern Apennines and Caucasus.

Craters such as Copernicus were born at a relatively late stage, perhaps even less than one æon ago, by which time the crust had become so firm that the magma was mainly trapped beneath it. Gradually the activity died away, until at last crater formation ceased altogether—apart from the impact structures which had been produced here and there all through lunar history, and which are presumably being produced occasionally even today.

I am well aware that this interpretation is open to challenge, but it does seem to fit the facts, and it can also explain the marked difference the Earth-turned and the far hemisphere of the Moon. We can see, too, why craters which break into later formations have not ruined the walls of their victims, and the almost invariable overlapping rule of smaller craters disturbing larger ones becomes logical. Seas are put into their proper class, either as outsize craters (Imbrium, Orientale, Crisium) or as overflows into former basins (Procellarum, Frigoris, Australe).

There are, of course, variations on the main theme. One of them is due to Dr. Gilbert Fielder, probably Britain's leading lunar authority at the present time. Fielder was originally an impact supporter, but later changed his mind, and now believes that the main structures are endogenic. Generally we are in agreement, but we differ in one respect. I believe that ghost craters, such as Stadius in the area of Copernicus, are ancient; Fielder regards them as young by lunar standards, and thinks that the growth of a crater is a very slow process indeed, so that Stadius and others of its type are still evolving. So far as the main controversy is concerned, he places great faith in his statistical studies, which do not agree with the idea of random impact, and he has made a close study of the grid system, which is more complicated than might be thought.

I repeat that 'vulcanism', in the sense that I have been using the term, covers many different processes, and some fascinating experiments by Dr. Allan Mills at Leicester may provide a better solution to the problem than anything which has been proposed before. Mills' theory involves what is called a fluidized bed. If a stream of gas is introduced below particular material, and the flow is then increased, the bed will expand, and behave as if it is liquid. Bubbles will appear, and finally material will be blown out of the vent; different rates and pressures in the gas-streams can produce craters strikingly similar to those of the Moon. This differs from the caldera picture, but it still attributes the craters to internal rather than external action. The experimental results are spectacular, and the whole chain or argument is so clear and logical that it rings very true.

All in all, our main task now is to decide which process has played the major rôle: bombardment, or internal action in some form or other. I have outlined the problem in the thoroughly biased account given above; reading a different book will give a completely different view. There is no harm in this. Only by reasoned argument will the truth finally emerge, and I repeat that I am very ready to believe that all my own ideas are very wide of the mark. Meanwhile, what can we learn from other worlds?

Here, of course, we rely entirely upon the space-probe results. Mars was the first world beyond the Moon to be found to have craters on its surface. The first indications came from Mariner 4, which by-passed the planet in 1965; better results were obtained from Mariners 6 and 7 in 1969, but most of our detailed knowledge comes from one vehicle—Mariner 9, which was put into a path round Mars at the end of 1971, and which continued to send back high-quality pictures for several months before its power finally gave out. The Martian scene proved to be utterly unlike what most people had expected. Instead of gentle, rolling plains, there were mountains, valleys, craters—and volcanoes. There is not the slightest doubt that some of the volcanoes are similar in type to those of our own Hawaiian Islands, though they are much larger and more massive, the Olympus Mons, for instance, towers to 15 miles above the general level of the surface, and has a 300-mile base, while the summit caldera has a diameter of 40 miles. Associated with the volcanoes are canyons which look as though they have been water-cut, though the atmospheric pressure at the surface of Mars today (below 10 millibars) is much too low for any liquid water to exist there.

Craters abound, and look superficially like those of the Moon, though there are significant differences in detail.

Beyond the path of Mars lies the zone of asteroids or minor planets. As we have noted, there is probably no essential difference between a small asteroid and a large meteorite, and therefore Mars, much closer to the main swarm than we are, would be expected to be more thickly cratered if impacts were responsible. Actually the crater frequency is much lower, but it is fair to say that erosion may have played an important part, since even though the atmosphere of Mars is thin it is certainly not negligible.

Yet this leads us into fresh difficulties. If the Martian craters and valleys were ancient by geological standards (that is to say, more than a few tens of thousands of years old) they would presumably be partly dust-filled, since dust-storms are not uncommon on Mars and can be easily observed from the Earth. But the craters are well-defined and lacking in any sort of dusty filling; the same applies to the canyons, which form vast drainage systems associated with the principal volcanoes. It may be that Mars experiences periodical active spells, when new craters are formed and gases are sent out from the volcanic vents in sufficient quantity to cause a temporary thickening of the atmosphere. This would explain the lack of erosion—which is hard to account for if the craters are impact structures, since they would then, inevitably, be very ancient indeed. This is no place to delve into Martian topics,* but we may be sure that vulcanism has operated there on a Titanic scale.

Mariner 9 also sent back photographs of Phobos and Deimos, the two midget Martian satellites. They, too, are cratered. The pits may be due either to impact or to what we may term blowhole activity; there is no firm evidence either way. Suggestions have been made that if a dwarf world such as Phobos, with a longer diameter of less than 20 miles, were battered by missiles capable of producing craters as large as those actually seen, the whole satellite would have been broken up; the biggest crater, now named Stickney, is 4 miles across. However, all this is highly speculative, if only because we know nothing definite about the density or rigidity of either Phobos or Deimos.

Venus is definitely a problem planet. It is about the same

* See *Mars*, by C. A. Cross and myself (Mitchell Beazley, London 1974).

size as the Earth, and dense clouds hide its surface completely. The surface temperature is of the order of 900 degrees Fahrenheit, and the atmosphere is mainly carbon dioxide. Since, moreover, the clouds seem to contain a great deal of sulphuric acid, Venus is not the sort of world likely to harbour life of our kind. We can never see the surface; but in 1973 radar measurements from America proved that craters exist there too. Examination of limited areas revealed that there are large, shallow craters, some of them more than 100 miles across. The impact theory is in serious trouble here. It would need a very large missile to plunge intact through the dense atmosphere and produce a surface crater.

Further developments occurred in October 1975, when the Russians managed to soft-land two probes (Veneras 9 and 10) and obtain pictures direct from the surface. The views were quite extraordinary: rounded rocks, with finer 'soil' in between, and no craters on view. Yet the range was very restricted, and neither probe managed to operate for much more than an hour before being put permanently out of action by the immensely hostile conditions. There were two extra surprises: there was more light than had been expected (about the same as the light-level at noon in Moscow on a cloudy day, as the Soviet authorities commented), and the wind-speed was a mere 7 knots or so. This, in an atmosphere as dense as that of Venus, is still as violent as the force of waves breaking against a cliff during a 50-knot gale on Earth; but it is less than had been anticipated, so that surface erosion may also be less. Certainly Venus has many problems in store for us, but in the present context it is probably enough to say only that there are craters there, even though their forms may well be different from those of the Earth, Moon or Mars.

Finally there is Mercury, photographed from close range in 1974 and 1975 from Mariner 10, the first two-planet space-craft. Here we have a scene which is undeniably lunar. There is a huge basin, known as the Caloris Basin; there are craters with central peaks, and craters without; there are valleys, ridges and ray centres—in fact, everything which we can see on the Moon, although there are fewer of the large plains of the Mare Imbrium type. The general law of non-random distribution seems to apply. It is hardly likely that meteorites would exist in large numbers so close to the Sun as Mercury, but here again we are wandering into speculation, though it is logical to suppose that the crater-producing forces on Mercury and the Moon have been the same.

There we must leave the whole problem for the moment. Volcanic rocks on the Moon are widespread, but whether the craters are of internal origin or were produced by bombardment is quite another matter. Despite Apollo, the 'Hundred Years' War' is still far from over.

Chapter Fifteen

FLASHES, GLOWS AND MOONQUAKES

'THE MOON IS A CHANGELESS WORLD.' This was the view of Beer and Mädler, the greatest of all pioneer lunar observers, and it was also the official edict until very recently. Nowadays we know that things are not so straightforward as this, and that the Moon is not totally inert. Yet what about major structural alterations?

There have been several cases of reported changes on the Moon: disappearing craters, newly-formed craters and so on. I must preface my comments by admitting that I have no faith in any changes of this kind; but it is always interesting to delve into history, and a good place to begin is at Linné, on the Mare Serenitatis. Linné is one of the most-studied objects on the whole Moon, and observers have every reason to be grateful to it, since it was the direct cause of the reawakening of interest in selenography from 1866 onward.

Lohrmann, in 1834, looked at Linné and described it as 'the second most conspicuous crater on the plain . . . it has a diameter of about 6 miles,* is very deep, and can be seen under all angles of illumination'. Mädler, at about the same time, wrote: 'The deepness of the crater must be considerable, for I have found an interior shadow when the Sun has attained 30 degrees. I have never seen a central mountain on the floor.' Both observers drew it, measured it and used it as a reference point. It also appears as a conspicuous crater on six drawings made by Julius Schmidt between 1841 and 1843.

All this seemed definite enough. Yet on 16 October, 1866, Schmidt was examining the Mare Serenitatis when he suddenly realized that Linné had disappeared. Where the old crater had stood, all that remained was a small whitish patch. It was a startling discovery, but Schmidt had no doubts about it.

---

* Lohrmann actually said 'somewhat more than 1 mile', but the old German mile is equal to 4½ of ours. Mädler gave 1·4 German miles, thus agreeing with Lohrmann.

His announcement caused a tremendous sensation. Up to then, Mädler's view of the Moon as a dead, changeless world had been accepted without question for many years, and astronomers were not inclined to change their opinions. Hundreds of telescopes were pointed at Linné, and many drawings were made of it (photography was still at an early stage). The results were not in good agreement, but at least it was clear that the deep crater described by the old observers had gone—if, of course, it had ever existed. In its place was a whitish patch, containing a tiny feature which was sometimes described as a craterlet and sometimes as a hill. One particularly famous observer, Angelo Secchi—a pioneer of stellar spectroscopy—looked at it on 11 February, 1867, using the powerful telescope at the Vatican Observatory, and wrote that 'there is no doubt that a change has occurred'. Sir John Herschel suggested that a moonquake had shaken down the walls of the old crater, and that the hollow had been filled with rising lava.

Probably the leading active selenographer of the time was Edmund Neison, author of the classic book published in 1875. I have found an article by him written for the *Quarterly Journal of Science* for January 1877, and it seems worth quoting:

According to three or more independent selengraphers, the most experienced and eminent that Science has seen, the object named Linné was a conspicuous crater of large diameter and depth. Now in its place all that exists is a tract of uneven ground, containing a small, scarcely-visible, insignificant crater-like object. It is impossible that the one could ever be systematically mistaken for the other. It is inconceivable how our three greatest selenographers could have systematically and independently made the same blunder, and that one blunder only. . . . A real physical change on the Moon's surface must therefore have occurred at this point.

Of the 'three greatest selenographers' (Lohrmann, Beer and Mädler) only Mädler remained alive in 1866. He accepted the evidence of change, but he nevertheless wrote that when he observed Linné in May 1867 he 'found it shaped exactly, and with the same throw of shadow, as I remember to have seen it in 1831. The event, of whatever nature it may have been, must have passed away without leaving any trace observable by me.'

If Linné showed no change to Mädler, the one man who should have been in a position to know, the evidence in favour of a genuine alteration is weakened at once. Moreover, it is true that the testimony depends only upon Schmidt and Lohrmann, apart from Mädler himself (Beer, co-author of the great map, apparently did not do very much of the actual observing)—and before 1843, when he saw Linné as a crater, Schmidt was still young and inexperienced. Earlier charts are not helpful. Cassini's map of 1692 shows a feature in the right position, but the chart is very rough, and the only surviving sketch of the area by Schröter was made to show the bright rays which cross the Mare Serenitatis; the other details are merely put in for position, and both Linné and Bessel are shown as spots, so that the drawing cannot be said to prove anything either way.

There is also the point that Linné can today appear as a crater even in a small telescope when seen in a modest telescope —as I first saw for myself (to my surprise, I confess) on 23 March, 1961, when I was using an $8\frac{1}{2}$-inch reflector. And, of course, we now have the close-range probe pictures, which show Linné as it really is; a clear-cut small crater, obviously young by lunar standards and surrounded by a light area. This at least could well be an impact structure. Weighing the pros and cons, I am of the firm opinion that no change has occurred there, and by now I doubt whether many people will disagree.

I have spent some time in discussing Linné because it has always been regarded as the one instance of alleged change which is worthy of serious consideration. The other 'lunar alterations' are even less convincing, so that I will pass over them more briefly. Schröter drew a 'large distinct crater, with bright walls and a dusky floor' on the eastern border of the Mare Crisium, and named it Alhazen; Mädler could not find it, and transferred the name to another formation. Mädler himself showed a five-mile crater near Alpetragius which had similarly 'gone missing' later on when Schmidt looked for it; on the other hand the large walled plain Cleomedes contains a small crater considered by Schröter to have been formed about October 1789, since he had missed it previously but saw it quite clearly afterwards. Judging from the look of the areas concerned, any alterations there seem wildly improbable, and errors in observation give a much more likely answer. Remember that before 1866 (the 'Linné' period) there were only a very few observers looking seriously at the Moon, and Mädler, the best of them, used a refractor of only 3¾ inches aperture for his main lunar work.

There was also the case of Hyginus N, a rimless depression close to the famous cleft, which was first seen on 27 May, 1878 by the German observer H. Klein. It was, Klein reported, three miles in diameter, and filled with shadow under oblique lighting; as he had never seen it before he thought that it must be new. I am sure we can discount any change here; I mention it only because it caused a great deal of discussion at the time, and for years afterwards.

On the Mare Foecunditatis there are two small craters, Messier and Messier A, from which extend a double ray which gives the pair a strange resemblance to a comet. Beer and Mädler made over three hundred drawings of them, between 1829 and 1837, and described them as being exactly alike. 'To the west* of Messier there appears an identical formation. Diameter, shape, height and depth, colour of interior are the same, and even the positions of the peaks; everything points to the fact that we have here either a most remarkable coincidence, or that some as yet unknown law of nature has been at work.' Today, Messier and A generally appear different in both shape and in size, A being the larger and less regular; but the angle of sunlight striking them is all-important here, and there is no doubt at all that the alterations are purely optical. Under some illuminations they really do look alike. (H. H. Nininger, the American meteorite expert, suggested in 1952 that Messier and its twin had been formed by a meteorite which ploughed its way through a ridge! The idea is intriguing, but not, I fear, in accord with the facts.)

Finally, there is the structure to which I gave the unofficial name of Mädler's Square. It lies closely east of the crater Fontenelle, on the edge of the Mare Frigoris. Mädler drew a regular square enclosure with high mountainous walls, and in

---

* Mädler actually said 'east of Messier'; here, as in other descriptions of early work, I have altered east and west so as to conform with the IAU reversal. On some maps, too, Messier A was named 'W. H. Pickering', a name which has been dropped from the latest IAU official list.

1876 Neison wrote that it was 'a perfect square, enclosed by long, straight walls about 65 miles in length and 1 mile in breadth, from 250 to 300 feet in height'. Actually the enclosure is incomplete, and the walls are anything but regular. The illusion is caused by the slightly darker tone of the interior of the 'square', and there has certainly been no change; Schröter, twenty years before Mädler, drew the area exactly as it is at present.

So much for alleged structural alterations on the Moon. Nowadays they do not occur; no new craters are being formed, apart possibly from occasional small impact pits, and to all intents and purposes the lunar surface looks the same as it must have done millions of years ago. This is one argument which has been settled, and is now of historical interest only; possibly I have spent too much time on it. But when we come to short-lived, transient flashes, obscurations and glows, the situation is very different, and we come up to modern times.

Occasional flashes have been reported, and have been explained as being due to meteorite falls. It was once calculated that a 10-lb. meteorite should produce a flash bright enough to be seen with the naked eye from Earth, but nothing of this sort has ever been seen. I have collected and analyzed most of the flash reports, but to be candid there is only one which impresses me. This was an observation made on 15 April, 1948 by F. H. Thornton, a very experienced observer who was using a 9-inch reflector under excellent conditions. He wrote as follows:

When I was examining Plato, I saw at its eastern rim, just inside the wall, a minute but brilliant flash of light. The nearest approach to a description of this is to say that it resembled that flash of an AA shell exploding in the air at a distance of about ten miles. In colour it was on the orange side of yellow. . . . My first thought was that it was due to a large fall of rock, but I changed my opinion when I realized that close as it seemed to be to the mountain wall, it was possibly over half a mile away.

Was this a meteoritic fall? It certainly cannot have been anything in the nature of a volcanic outbreak, because it was so brief. I am, I admit, influenced by the fact that I have a great admiration for Thornton's skill and judgement, but there are no other flash reports that I take very seriously. So let us move on to what I have christened TLP, or Transient Lunar Phenomena—a term which now seems to have become generally accepted.

Lunar features change rapidly under the influence of altering illumination. A feature which shows up as a well-marked crater one evening may look like a nebulous patch the next, and there are even cases in which the actual forms appear to vary, Messier being the best example. Yet in general all the details on the Moon are sharp and clear-cut, as is only to be expected upon a world which lacks atmosphere. Local obscurations, blurring the detail over a restricted area, could not be explained as 'clouds' or 'mists', but so many reports of them have come in over the years that there can be little doubt of their reality. They include glows, usually reddish in colour; colourless obscurations over limited regions, and—occasionally—larger obscurations which do not hide the surface details, but merely make them less distinct.

In the days when it was still thought that the Moon had an atmosphere, albeit a tenuous one, some of the phenomena were put down to what might be called lunar twilight. Comparisons were made with Venus, which has of course a very dense atmosphere rising high above the surface of the planet. When Venus is at the crescent stage, the horns are often extended, and pioneer lunar observers, including Schröter and Mädler, described similar prolongations at the horns of the crescent Moon. Later reports were also made, but I am frankly sceptical, partly—I admit—because my own searches have been so completely fruitless, but mainly because no reliable reports of horn prolongation have come in since the start of the Space Age. If accurate, they would be very hard to explain, but I am sure that they can be put down to errors in interpretation.

Localized obscurations inside or near walled formations come into a different category, and records of them were not confined to amateur astronomers. Thus in 1892 Edward Emerson Barnard, the great American observer who was noted for his keen eyes (for instance, he discovered the fifth satellite of Jupiter), recorded that on one occasion the bright ray-crater Thales was 'filled with pale luminous haze', though the surrounding features were perfectly sharp and clear-cut. In 1902 the French astronomer Charbonneaux, using one of the world's largest refractors (the Meudon 33-inch, at the Paris Observatory) described how he saw a small but unmistakable white 'cloud' form close to Theætetus, in the region of the Apennines, and various localized obscurations were reported by W. H. Pickering.

Variations were reported, too, inside the 60-mile, dark-floored Plato, one of the most-studied formations on the Moon. The craterlets on the floor are sometimes invisible when they should be obvious. I can cite a personal case here. Before midnight on 3 April, 1952 I was unable to see them at all, though I too was using the Meudon refractor under good conditions; four hours later T. A. Cragg, now of the Mount Wilson Observatory, looked at Plato with a 12½-inch reflector and saw that the floor looked blank. He was in no doubt that some local obscuration was responsible. Plato has a long history of similar anomalies.

Despite the professional observations, it was true that the bulk of the reports came from amateurs, and the local obscurations were officially dismissed as being due to imagination or tricks of the eye. Those who believed otherwise, such as myself, were not taken very seriously. This was understandable enough, because to see one genuine event means many hours of fruitless checking—something which no professional, busy with more important matters, has the time to do. So the amateurs, notably those of the Lunar Section of the British Astronomical Association, went on with their patient work and waited to see what would emerge from it.

The whole situation was transformed by one episode, in which I played a minor and totally undistinguished part. During 1955 Dr. Dinsmore Alter, using the 60-inch reflector at Mount Wilson, took some photographs of the two large walled plains Alphonsus and Arzachel, which are members of the Ptolemæus chain, and which lie near the centre of the Moon's disk as seen from Earth. Alter took pictures in both infra-red light and in blue-violet. As almost everyone knows, infra-red will penetrate haze or mist, while light of shorter wavelength will be blocked or scattered by it. (This is why our sky is blue; the shorter wavelengths of the Sun's radiation are spread around, while the longer wavelengths are not—at least, not to the same extent.) Alter naturally expected that in his sets of photographs Alphonsus and Arzachel would be similarly clear-cut, but he found that on several occasions part of the floor of Alphonsus was blurred in the blue-violet pictures. He wrote: 'For some reason the blue-violet photographs lose more detail in the east side of Alphonsus than they do in the floor of Arzachel. This is not true of the infra-red ones. . . . There is a temptation to interpret these results immediately as being due to a thin atmosphere, either temporary or permanent, over the floor of Alphonsus. The theoretical difficulties inherent in such a hypothesis are, however, strong enough to forbid whole-hearted acceptance of it.'

This was by no means the first time that local effects had been reported in Alphonsus. I had some correspondence with Alter and various other professional astronomers, including N. A. Kozyrev, of the Crimean Astrophysical Observatory in the U.S.S.R. I suggested that it would be worth while to keep a careful watch on the area, and Kozyrev was among those who did so, using the 50-inch reflector at the Crimea. His method was to take regular spectrograms (that is to say, photographic spectra), and since the Crimean reflector has no separate guiding telescope he had to watch during the exposure time to make sure that there was no drift of the image. While doing this, at 0100 hours G.M.T. on 3 November, 1958, he noticed that the central peak of Alphonsus had become blurred, and was apparently engulfed in a reddish 'cloud'. Another spectrogram, taken between 0300 and 0330 hours G.M.T., proved to be remarkably interesting. While guiding the telescope, Kozyrev kept his eyes on Alphonsus, and noticed that the central peak had become abnormally bright. Suddenly the brilliance began to fade; Kozyrev immediately stopped the exposure and started a new one, which was completed at 0345 G.M.T. By the time it was finished, everything was normal once more, and Alphonsus looked the same as it usually does.

The announcement of Kozyrev's results took many astronomers by surprise. Kozyrev himself was quite definite: 'On the spectrogram recorded on 3 November, 1 hour U.T., the central peak of the crater appears redder than normal. Probably at this time the peak was being observed and illuminated by the Sun through the dust (ashes) being thrown up by the eruption.' In a letter to me written shortly afterwards, he said that the spectrograms showed that 'hot carbon gas had been sent out, causing a rise in temperature of perhaps 2,000 degrees'.

The hot carbon gas and the marked rise in temperature were questioned by other theorists, but at least one thing was evident: there had been a disturbance of some kind. 'Eruption' is a misleading term, but certainly what Kozyrev had seen was a TLP. Checking the records, I recalled that as long ago as 1882 a well-known German selenographer, H. Klein, had claimed that volcanic phenomena were still going on inside Alphonsus, though I would again stress the danger of putting too much faith in last-century reports.

The next step was to see if there had been any permanent change in the area. During the following weeks and months several observers reported red patches near the site, and these were interpreted as being due to coloured material thrown out at the time of the disturbance; for instance Brian Warner, now Professor of Astronomy at the University of Cape Town but then working at the University of London Observatory, used the 18-inch refractor there, and described the patch as 'bright red'. I was less successful, and I admit that I was never able to see any red patch at all; it certainly seems to be absent now, but the 1958-9 reports are not at all easy to explain away.

Since then there have been other records of red TLP in the crater, plus one more spectrographic observation by Kozyrev on 23 October, 1959, though on that occasion nothing was seen visually. The floor has been photographed from the Orbiters, and of course Ranger 9 actually landed there. There can be little serious doubt of extensive past vulcanism. Kozyrev maintains that the central peak of Alphonsus is a genuine volcano with a funnel, but I admit to having grave doubts, largely because I question whether the Moon has sufficient internal energy at the present time to produce a bona-fide eruption. Neither can I accept a temperature-rise of thousands of degrees —or, indeed, any increased heat at all. E. J. Öpik questioned the reality of the phenomenon as a true disturbance, and attributed it to fluorescence, but recent work has shown that no fluorescent effects on the Moon can be strong enough to cause glows visible from Earth.

The next really major development came in 1963, when, on

30 October, J. Greenacre and E. Barr, at the Lowell Observatory in Arizona, observed colour in the region of Aristarchus. The phenomena included red and pink patches, and were quite unmistakable. In the following month similar events were seen, and were confirmed by P. Boyce at the Perkins Observatory, using a 69-inch reflector—a giant telescope by any standards. Detailed accounts were published; and if any lingering doubts about the reality of TLP remained after the Kozyrev episode, they were finally dispelled.

Aristarchus is the brightest crater on the Moon. It is only 23 miles in diameter and 6,000 feet deep, but it is so brilliant that it is always instantly recognizable even when lit only by earthshine; in 1787 no less a person that Sir William Herschel mistook it for a volcano in eruption. It has a central mountain; the walls are terraced, and are crossed by dark bands which are easy objects. The bands are due to differences in level and in surface texture, as we now know from the Orbiter and Apollo photographs. Close beside Aristarchus is Herodotus, of similar size but with a darkish floor, and from Herodotus extends the great valley often known as Schröter's Valley in honour of its discoverer. Aristarchus is the most event-prone crater on the Moon, and it is responsible for more than half the total number of reported TLP. Gaseous emissions from it have been confirmed spectrographically, and on 19 July, 1969 activity was seen by the astronauts of Apollo 11, who were then together in the command module orbiting the Moon. Armstrong, Aldrin and Collins used binoculars, and reported a luminous northwest wall, 'more active' than anywhere else on the surface. Added confirmation came from observers on Earth, who reported TLP in Aristarchus at and around that time.*

To run ahead of the story: there were further developments with the flight of Apollo 15, in 1971, when the command module carried a special device intended to detect what are known as alpha particles. These are produced by the decay of the radioactive gas radon, which in turn comes from uranium and thorium. If radon gas diffuses through the regolith, it will release atoms into the tenuous lunar atmo-

---

* Unfortunately I could take no part in the observational programme; I was in the B.B.C. television studio throughout the mission, carrying out 'live' commentaries.

sphere; when these atoms decay, alpha particles will be emitted. As Apollo 15 passed seventy miles above Aristarchus, there was a significant rise in the numbers of alpha particles emitted by radon-222 (that is to say, the radon isotope with an atomic weight of 222). Particles associated with the decay of the other main kind of radon, radon-220, were not found, and this showed that the effect was not due merely to a local excess on the surface of uranium and thorium. The radon isotopes must have diffused through the regolith from below. Radon-222 could be (and was) detected, but radon-220 atoms have a very brief existence; they last for less than one minute, and would not persist for long enough to come through the regolith. Therefore, there seems no doubt that radioactivity in the Aristarchus area is responsible, and the scientists who carried out the analyses—P. Gorenstein and P. Bjorkholm—added that: 'The observed radon emanation is associated with the same internal processes which will on occasion emit volatiles in sufficient quantity to produce observable optical effects.'

There is a clear link here with TLP, and it is worth noting that similar radon emissions have been found in the region of Grimaldi, another event-prone area. At the time of the Lowell observations, in 1963, Apollo lay in the future, but the whole problem of 'lunar events' was taken up energetically, and the patient amateurs were at last joined by professional astronomers. One method of detecting coloured phenomena involves the use of rotating filters, and the Lunar Section of the British Astronomical Association introduced what has become known as the 'Moonblink' device, described in the Appendix. Various positive reports came in, notably that of 30 April, 1966, when a red TLP in the area of the walled plain Gassendi was seen by several completely independent observers. This was, in fact, the most unmistakable red event that I have ever seen on the Moon, and it persisted for about four hours. The main feature was a wedge-shaped, reddish-orange streak extending from the wall of Gassendi right across to the central peak.

In America, Barbara Middlehurst and her colleagues at the University of Arizona and the Goddard Space Flight Center began compiling a catalogue of all lunar events which had been reported since the start of serious telescopic observation of the Moon. I was doing the same thing independently at Armagh

Observatory, and when we compared notes we found that our results were similar, so that the catalogue was eventually published under our joint authorship. It took the story up to 1967, and there were 579 reports altogether, though I would be the last to claim that all are reliable (in fact, I am quite certain that they are not!). In 1971 I extended the catalogue, bringing it up to date; the last entry was No. 713.

When we began to analyse the reports, some curious facts emerged. The event sites were not distributed at random, but lay around the edges of the regular maria and in regions rich in rills. The chart given here (Fig. 47), drawn up in 1975, shows all the areas in which events have been reliably reported; red TLP are starred, others indicated by black dots. The chart speaks for itself. For instance, the boundaries of the Mare Imbrium and the Mare Serenitatis are clearly outlined by event-prone areas, and so is the smaller Mare Crisium. I have also found that most of the red events are seen in the western hemisphere, while the others are concentrated in the Mare Tranquillitatis area to the east, so that different areas on the Moon seem to produce different kinds of events.

One has always to be on guard against observational bias. Because Aristarchus is so notorious as an event area it is closely watched, and a TLP there is more likely to be noticed than one in—say—Petavius, where no events have ever been reported. Also, I have to admit to being sceptical about some of the entries in the American catalogues, because, in my view, it is dangerous to trust anything except reports made by observers of proved skill and using adequate telescopes. Even then, independent confirmation is—to put it mildly—highly desirable.

What, then, causes these strange, elusive phenomena? I have an utter lack of faith in the idea of volcanic eruptions of any sort; the Moon is simply not that kind of world nowadays. Fluorescence, as we have noted, was seriously considered, but recent work carried out by Dr. J. Geake at Manchester University has shown that it cannot possibly be the answer. The theory of gas release from below the lunar crust has much more to recommend it, and there are various reasons for believing this to be the right explanation.

Professor S. Miyamoto of the Kwasan Observatory in Japan

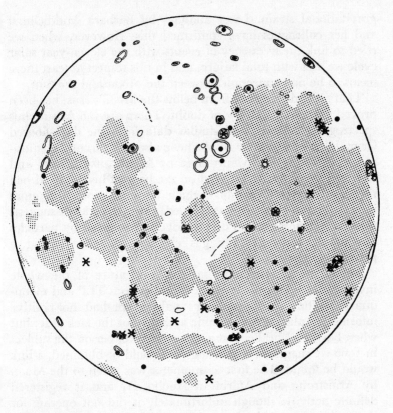

✳ Red events
● Others

Fig. 47. Distribution of Lunar Events, as drawn up from the catalogues by Barbara Middlehurst and myself. Stars indicate red events: dots, all others. Only areas in which events have been reliably recorded are shown

made the first really useful comment when, in 1960, he considered the possibility of gas release, and pointed out that in such a case the events would be more frequent in areas rich in rills, which go down for some way into the crust. Miyamoto also wondered whether the Earth's gravitational pull could produce any observable effects. Then, in 1963, Dr. Jack Green in America analyzed the records available and found that TLP are commonest near lunar perigee, when the Moon is at its closest to the Earth and its surface is under maximum

209

gravitational strain. Later analyses by Barbara Middlehurst and her colleagues have confirmed this. However, when we tried to link the frequency of events with the eleven-year solar cycle we met with total failure, and in this respect at least there seems to be no connection between the Moon and the Sun.

That gases do seep out from below the Moon's crust has been proved without a shadow of doubt. There are the Aristarchus spectrogram records, and similar data from the dark-floored Grimaldi, while positive results have also come from the equipment left in the Fra Mauro area by Astronauts Shepard and Mitchell from Apollo 14. And on the basis of his fluidized bed theory of crater formation (which becomes more and more impressive the closer it is examined), Dr. Allan Mills suggests that the phenomena are due to friction produced by particles moving around after the gas ejection, thereby making a perceptible glow.

Yet all in all, the most satisfactory feature of the whole investigation so far is the connection between TLP and moonquakes. When we began our analyses we had no positive information about lunar seismic activity—or the lack of it; but when Barbara Middlehurst and I wrote a paper on the subject in 1966 we suggested that when data could be obtained, a link would be found. The first seismometer was taken to the Moon by Armstrong and Aldrin in Apollo 11, and it registered definite activity, though unfortunately it did not operate for long. Since then seismic equipment has been set up by the astronauts of Apollos 12 and 14 to 17 inclusive, and all this equipment is still working. Moonquakes do occur, and are very frequent. It is from them that the present picture of the Moon's structure has been drawn up—a 35-mile crust, below which is a mantle extending down to 600 miles, below which again are the asthenosphere and finally the molten core, at a depth of about 900 miles.

The moonquakes seem to be of two kinds (Fig. 48). The shallow ones occur at depths of from 15 to 150 miles, and the deep ones at up to over 600 miles, with a peak frequency at just less than 600 miles. In other words, moonquakes occur in a region half-way between the lunar surface and the core, though in the region between 180 and 500 miles below ground level very few shocks are recorded. Each seismic station records one

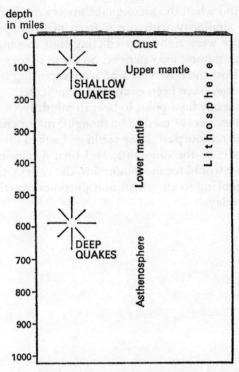

Fig. 48. Depths of Moonquakes

or two moonquakes daily, though I hasten to add that by terrestrial standards they are feeble. It has been said that an observer on the Moon standing right above the focal point of an earthquake would hardly notice anything at all, so that there is no danger to a future Lunar Base on this score. From the seismic point of view, the Mare Imbrium is far safer than Tokyo or San Francisco. Because the Moon is so relatively 'quiet' seismically, the recording devices can be made a thousand times more sensitive than is practical on the turbulent Earth.

There is no doubt that the moonquakes are genuine, and that they are produced by the same sort of mechanism as our own shocks. (Meteorite impacts have also been recorded by the lunar seismometers, but luckily they produce different kinds of wave-patterns, so that they can be weeded out from the analyses.) And when the moonquake areas are plotted, there is an obvious similarity with the pattern of event-prone areas. Moreover, we were right in predicting that moonquakes, like TLP, are commonest near perigee. So far as it goes, everything fits, and at least the observers who were reporting lunar events many years ago have been completely vindicated.

Yet there is one final point to bear in mind. Even though the Moon is not so inert as used to be thought, major changes there belong to the remote past. If we could go back to the days of the cavemen, or even the dinosaurs, and turn a telescope toward the Moon, we would see the mountains, the valleys, the seas and the craters looking to all intents and purposes exactly the same as they do today.

Chapter Sixteen

BEYOND APOLLO

When Eugene Cernan re-entered the lunar module of Apollo 17, and joined Jack Schmitt inside the vehicle, the first phase of the exploration of the Moon had come to an end. Much had been learned. The Moon had been explored, at least in half a dozen localized sites; it had been surveyed in detail from orbit, and many samples of its regolith had been collected. Various 'bogeys' had also been laid; for instance there seemed to be no danger from meteoritic bombardment, and the astronauts had no difficulty whatsoever in walking around, despite their cumbersome space-suits.

Yet Apollo, as I have stressed, was a reconnaissance programme, and could never be anything more. It has to be considered jointly with the Russian plan of exploring by using automatic probes, which was of immense value even though so much less spectacular. There are as yet no definite schedules for revisiting the Moon, and I am sure that we must await the development of better (and cheaper!) vehicles. The space-shuttle, now being developed as quickly as possible, is an essential step.

Of course, I can speak with some confidence only about the United States plans, because the Russians are not generally prone to make announcements in advance. It is quite possible that they will send their first cosmonauts moonward well before the Americans are ready to recommence, in which case they might start by dumping supplies at a pre-selected site to await the arrival of the explorers. But on the whole it seems likely that they too will wait; and if I had to make a guess as to when the next men will go to the Moon, I would say 'around 1990'. The problems are not wholly, or even principally, scientific; there are economic and political pressures to be considered as well.

No doubt there will be many more lunar unmanned vehicles before 1990, and these will add to our store of knowledge, though I doubt whether they will contribute anything really revolutionary. Thereafter, the emphasis should be upon a manned Lunar Base, and we have to decide how far the Moon itself is likely to help us.

The main disappointment so far has been that we must now, I fear, abandon the idea of extracting water from the lunar rocks, for the simple reason that there is none to extract. Neither can we hope for any supplies of water in the form of subcrustal ice (an idea in which I personally never had any faith). Atmosphere, of course, is lacking, and there is no conceivable chance of our being able to produce one, which means that no earth-type organisms can survive in the open, unprotected. All in all, we must take the whole of our requirements with us: air, water, food—nothing can be found *in situ*. Obviously, then, it will be a long time before a lunar colony can become self-supporting.

Two favourite themes of science-fiction writers have always been the discovery on the Moon of valuable minerals, and the possibility of using the Moon as a military base. In fact, there is no evidence that the Moon contains anything of commercial value (diamonds, for instance, are conspicuous only by their absence), and in any case the transportation costs would make the whole idea uneconomical. So far as military operations are concerned I can only say, thankfully, that there are no prospects of anything of the kind. Even as a base for spying, the Moon is inferior to a satellite in orbit round the Earth; and the notion of building nuclear missiles there, and firing them across space to land on enemy countries, is as futile as the concept of building a vast mirror and focusing the Sun's heat on to one's opponents in the manner of a burning-glass. Neither is there any need for a 'military' Moon. If *homo sapiens* really wants to destroy itself, it can do so by using ground-to-ground missiles which have already been stockpiled.

On the other hand, the Moon will be ideal as a site for scientific research stations of all kinds. It will be an astronomer's paradise, with no inconvenient atmosphere to shut out the most important radiations, and with no interference from artificial lights; also, the low surface gravity will help in the building of bulky equipment (though there are some difficulties also, due to the behaviour of constructional materials in conditions of vacuum and great temperature-range). There is also the point that the sky seems to move around very gradually, because the Moon is a slow spinner, and this will be a help when long-exposure photographs of distant stars and galaxies are being taken; moreover the periods of darkness will be much longer. Incidentally, the north pole star of the Moon is not Polaris, but—at present—Zeta Draconis, though the polar point moves more quickly than ours, and describes a small circle in the sky over a period of eighteen years.

Radio astronomers will be equally happy, for much the same reasons. On the far side of the Moon, over which the Earth never rises, there will be complete 'radio quiet'. Not even pop music can penetrate to those isolated regions.

The physicist will welcome the chance to study all the radiations coming from space, and he will also have limitless hard vacuum, as well as numerous other disadvantages. Medical men will be able to carry out their researches under conditions of low gravity; it is not inconceivable that revolutionary surgical operations, impossible on Earth, will be practicable on the Moon. Even Big Business may benefit, because various chemical and mechanical processes may be helped by the lunar environment.

It is too early to say just what form a Lunar Base will take. The old pattern of plastic domes, kept inflated by the pressure of air inside them, may not be so very wide of the mark, though the early bases will certainly be much less regular and less impressive. At least we now know that there is no need to go underground to avoid being battered to pieces by meteorites, as was thought possible not so long ago.

If all goes well, the numbers of people on the Moon will increase steadily. Scientists will naturally predominate, but eventually there is no reason to doubt that others, too, will be able to make the journey. I doubt whether the fascinating idea of holidays on the Moon will become practicable for the next fifty years at least, but this, too, may come in time.

Finally, what about the international aspect? Here I may be something of an idealist. Probably the first bases will be either American or Russian, but they will be manned by scientists, who will be anxious to collaborate in the way that they manage to do, to a considerable extent, in the Antarctic. Once collaboration starts, it may spread to Earth—and so far as the Moon is concerned the politicians will be more or less powerless to interfere, because they will be unable to go there. It may not be absurd to suggest that in colonizing the Moon, we may go at least some way toward uniting the Earth.

Because I am myself essentially a lunar observer, this book has had an observational bias. I am well aware that much has been left out, and that many of the ideas I have put forward may prove to be wrong, but I hope that I have been able to give some impression of what the Moon is like. Strange, lifeless and hostile though it may be, it is a fascinating world, and today it has become of practical as well as of academic importance to us. It is within reach, and I have no doubt that some of those who now look upward at it will be able to go there. The Moon has set us a challenge; and if mankind chooses peace rather than war, it is a challenge which we will accept.

Appendix I

OBSERVING THE MOON

WHEN I WROTE the first edition of this book, more than a quarter of a century ago now, the Moon was still essentially the province of amateurs, and professional astronomers seldom looked at it systematically. The emphasis then was upon mapping. The libration regions, in particular, were poorly known, and it was important to chart them as well as possible—no easy matter, because of the great foreshortening. Today the whole situation has changed dramatically. Mapping has been completed, and the entire surface, apart from a tiny section near the lunar south pole, has been photographed in detail from the Orbiters and Apollos. Therefore, the rôle of the amateur is different.

After the first space-probes, and again after Apollo 11, there were many people who believed that further observation of the Moon from the Earth was scientifically a waste of time, which brings us back to the situation after the publication of the map by Beer and Mädler in 1838—when it was tacitly assumed that everything about the lunar surface was already known. In fact, however, lunar observation at the present time is as important as it has ever been, but the amateur worker has to be more specialized. In the notes which follow, I will try to outline some of the main lines of present-day research available to owners of modest telescopes. I do not propose to discuss more sophisticated equipment (photoelectric devices, etc.) because I am not competent to do so, and for the same reason I will give only a few brief comments about photography. First, then, there is the question of the minimum equipment needed.

(a) Telescopes and Procedure

One fact which is only too often forgotten is that nobody can carry out useful lunar work without a really good knowledge of the Moon's surface. Practical experience at the eye-end of a telescope is essential. My own method, which I recommend to all beginners, was to take an outline map upon which a couple of hundred craters were named, and then to spend my observing time in making at least two drawings of every named formation. One sketch would not suffice, because a crater (or any other feature) can show apparent changes according to the angle at which the sunlight strikes it, and these changes are very marked indeed, as I have stressed over and over again throughout this book. For instance, the huge walled plain Maginus is extremely prominent when near the terminator and filled with shadow, but at or near full moon, when there are virtually no shadows, Maginus is difficult to locate at all.

Full moon, moreover, is the very worst time to start observing, because the whole scene is dominated by the bright ray systems, and the disk appears as a confused medley of bright and darker patches. This was my first mistake, when I set out to learn my way around the Moon. I was nine years old, and the proud owner of a 3-inch refractor, which I still have (it had cost me £7. 10s., which seems ironical now!). I waited until full moon, looked at my map, and decided to make a start by drawing Ptolemæus, the great walled plain near the very centre of the disk. Not surprisingly, I could not find it. When I looked again during the next lunation, at half-moon, I identified it at once; indeed, I could hardly miss it.

Therefore, the procedure should be to start with the formations near the terminator. Draw them, and keep your results in a file; never throw them away. Next evening the terminator will be further advanced; draw some new craters, and re-draw the old ones. Repeat this procedure often enough, and you will soon become familiar with the lunar topography. It took me over a year to complete my first survey, and of course the drawings which I made were very rough and inaccurate, but to me they were invaluable. The programme is laborious, but it does work, and I have never been able to think of any short cut.

My 3-inch refractor was ideal for the purpose, and this is, in fact, the best possible telescope with which to begin. For 'learning' purposes an even smaller telescope is adequate, and it is far better to begin in a modest way, with a small telescope, than to use a powerful instrument at once.

Apart from occultation timings, which come into a different category, it is true that no practical lunar research can be carried out with a very small telescope. At the same time, it is my firm belief that such a telescope is essential if the observer is to turn into a really useful member of a team.

Inch for inch, a refractor is more effective than a reflector; thus a 3-inch refractor is useful, whereas a 3-inch reflector of conventional type is, to be honest, very limited. Probably the smallest aperture for the serious observer is 4 inches (for a refractor) or 6 inches (for a reflector), but, of course, the larger the telescope the greater the range. Thus for TLP observations I would not be happy with anything below $8\frac{1}{2}$ inches for a reflector, and even this would not satisfy me, though observers with keener eyes than mine will probably disagree. In any case, the owner of an $8\frac{1}{2}$-inch can play a full part in the programme. If he has, say, a $12\frac{1}{2}$-inch—well, so much the better.

Mountings are all-important. Trying to use a good telescope on a jelly-like mount is like trying to use a good record player with a worn needle, and it must be said, with regret, that even some professionally-made telescopes are fitted with inadequate stands. In selecting a telescope it is wise to take the utmost care, because it is the telescope which swallows up most of the initial expense of making a hobby out of astronomy. Fortunately, the expense is non-recurring.

Driving mechanisms, which compensate for the rotation of the Earth and keep the Moon (or whatever is being observed) in the field of view, are highly desirable, but they are not absolutely necessary except for astronomical photography. The visual observer can make do without them, though smooth manual slow motions are essential. I will not go into further detail now, because I have done so elsewhere.*

(b) Seeing and Magnification

Amateur observers have to face many problems. Trees, for instance, are a thorough nuisance, and generally not much can be done about them, so that a telescope too big to be portable has to be set up in the position which will give the best available view of the sky. Artificial lights are worse still, and town-dwellers are at a grave disadvantage; even those who (like myself) live far in the country have to contend with ever-increasing light pollution. It is hopeless to try to observe by poking a telescope through a window. Apart from the difficulty of keeping

* In *The Amateur Astronomer* (Lutterworth Press, 1974).

the telescope rigid, the temperature difference between the room and the outer air is bound to cause so much local atmospheric turbulence that the Moon will look as though it is shining through several layers of water. Sharpness of image is essential, and it can never be obtained from indoors.

Neither is it much good trying to observe a low Moon, as the light then reaching the observer is shining through a thick layer of atmosphere. Twilight, however, is no handicap on its own, and neither is slight mist harmful in the usual way, though even the thinnest layer of cloud is usually fatal. Often a very brilliant starlight night such as occurs after heavy rain will prove to be hopelessly unsteady, with the Moon's limb shimmering and rippling. Under such conditions there is nothing to be done except stop observing until things improve.

Beginners often make the mistake of using too high a power. On a really good night, of course, high magnifications may be used to advantage, but it is hopeless to put in a powerful eyepiece unless the resulting image is really sharp. Whenever a lower power will do equally well, it should be preferred. For a 3-inch refractor, a magnification of 100 is usually adequate; powers of 150 or so are useful only to finish off drawings which are already more or less complete. It is often said that the maximum usable magnification is 50 per inch of aperture, which would give 600 on a 12-inch telescope, and so on; but for most purposes a magnification of little more than 30 per inch is better. Of course, everything depends ultimately upon the conditions, the quality of the instrument, and the skill of the observer.

(c) Preliminary Drawings

For learning purposes, the first essential is to identify the crater concerned, which can be done from an outline map—though a photograph of the region will be invaluable as a check (remembering that the photograph will not generally show the same conditions of illumination and libration as those on the Moon at the time of observation).

When making a sketch of a lunar formation, the best plan is to start with a fairly low power, sketching in the main outlines—unless these have been prepared beforehand from a photograph, in which case allowance must be made for libration with a crater anywhere near the limb. Also indicate the shadows and coarser details. Then change to a higher magnification, and put in the fine details. If the night is really good, maximum possible power should be used to check each tiny feature, but details which are doubtful or suspected should be carefully distinguished. On the whole it is better to make a written note of doubtful objects rather than to show them on the actual drawing. Finally, re-check everything.

When the sketch is complete, add the following data: year, date, time (using the 24-hour clock, and never using Summer Time), telescope, magnification, name of observer, and seeing conditions. The scale usually favoured for seeing is that introduced by the Greek astronomer E. M. Antoniadi, ranging from 1 (perfect) down to 5 (very poor). If any of these notes are omitted, the drawing will promptly lose most of its value.

One term often encountered is 'colongitude'. This is equal numerically to the longitude of the morning terminator, measured westwards from the centre of the disk. Thus if the colongitude is 270 degrees, the morning terminator is at 90 degrees East—that is to say, the Moon is new. The colongitude is approximately 270, 000, 090 and 180 at new moon, first quarter, full moon and last quarter respectively. (At the morning terminator, the Sun is rising over that part of the Moon; at the evening terminator, it is setting.)

The most common of all faults is that of drawing too large an area on too small a scale. Twenty miles to the inch is a convenient guide, and it is better to be over-generous than parsimonious. I remember that on one occasion I was sent a sketch of the complete Mare Imbrium, made with a 5-inch refractor, in which the Mare itself was about four inches across and Plato perhaps a centimetre. Even if the sketch had been accurate (which it was not) it would still have been useless. Far better to select one crater or crater-group only, and make a drawing which is as faithful as possible.

Drawings can be made in various ways. Some observers manage to make representations which are artistic and accurate at the same time, whereas others with less skill—such as myself —prefer to keep to line sketches. From a scientific point of view, the main thing is to make the drawings accurate and easy to interpret.

(d) Photography

Lunar photography is a fascinating hobby, but since I have done so little of it myself I can only refer enthusiasts to some of the books listed in Appendix II. Putting an ordinary camera to the eyepiece of a small telescope and 'snapping' usually leads to disappointment, and for good pictures a driven telescope is necessary. Excellent results can be obtained with home-made equipment used in conjunction with, say, a 6-inch reflector, but the techniques involved are out of my own particular field.

(e) Research Programmes

Now that cartography is to all intents and purposes complete, drawings are generally made for the benefit of the observer himself only, but sometimes it is still helpful to make a sketch —for instance, with features which show unusual optical variations; Messier and its companion provide the classic case. The Lunar Section of the British Astronomical Association has also been carrying through a programme of measuring the depths and profiles of craters by estimating the amounts of shadow inside them at different times, though whether the method will turn out to be really useful remains to be seen.

TLP come into a different category, and here the visual observer is in his element, provided that he has an adequate telescope. First, make sure that you know the areas under study really well; the most promising sites are formations such as Alphonsus, Gassendi, Grimaldi and above all Aristarchus, though it would be a great mistake, and would distort the analyses, to concentrate upon these to the exclusion of all others. Really major events are visible in ordinary light, and it now seems that colourless TLP are commoner than red ones.

Use has also been made of the moonblink device, as developed in Britain by P. K. Sartory and V. A. Firsoff. This consists of a pair of filters, one red and one blue, which can be used in rapid succession. As the red filter will tend to suppress a red event and a blue filter will not, the well-known persistence of vision effect comes into play, and a faint red patch on the Moon may show up as a 'blinking spot'—hence the name. The device has its limitations, but it is remarkably sensitive, as experiments have shown. The drawing of my own moonblink (Fig. 49) will show the main construction. The eyepiece fits in at B; the

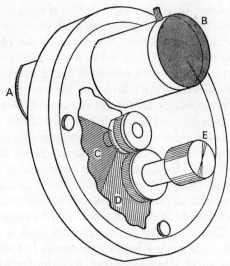

Fig. 49. A moonblink apparatus

filters are fitted in at C (red) and D (blue), with a clear area; the filters are rotated by turning the knob (E) and the device screws into the telescope in the ordinary way (A). Make sure that your filters are of the right type, and are of good optical quality; and of course the blink device cannot be used on a small telescope, as it involves too much loss of light.

TLP work is laborious, and many hours of fruitless searching will pass by before anything positive is seen. (Not so long ago I had a blank period of over two years; I hate to think of the period I spent searching during that time.) Moreover, the utmost care is needed. Many alleged events turn out to be spurious, caused by instrumental defects or by conditions in the atmosphere. Unfortunately, a wrong report will be worse than useless, because it will upset the analyses. If a suspected TLP is seen, check all the features in the surrounding area, and also all the way along the terminator. If other craters show the same effects, you may be sure that your TLP originates in the Earth's air rather than on the Moon.

The Lunar Section of the British Astronomical Association has organized a full TLP network which is proving very successful. Full details of the programme are given in the Section Handbook, listed in the bibliography.

Then, of course, there are occultations of stars by the Moon, which are still very useful provided that they are accurately timed. Occultation predictions are given in the Handbook of the British Astronomical Association for each year. All that is really needed is an adequate telescope plus a good stop-watch; of course the observer must also know his own position on the Earth very accurately. Usually the occultation is virtually instantaneous, though, as I have already said, there are cases of fading immersions—usually (or always?) produced when the occulted star is a close binary. If a star by-passes the upper or lower limb of the Moon, it may be momentarily hidden by mountain peaks along the limb, and these grazing occultations are of special importance, though nearly always the observer has to take a portable telescope to the extremely narrow critical line. One advantage of occultation work is that for the brighter stars, at least, it is within the range of the owner of a comparatively small telescope. Here, too, the Lunar Section of the British Astronomical Association has a wide and effective network of observers.

Appendix II

LUNAR LITERATURE AND LUNAR MAPS

NOWADAYS THE LITERATURE of the Moon is vast, and all I have done here is to select some of the books which I have myself found particularly useful.

First and, in my view, foremost is *The Amateur Astronomer's Photographic Lunar Atlas*, by Henry Hatfield (Lutterworth Press). This consists essentially of a series of photographs, covering the Moon under various conditions of illumination, and each accompanied by a facing map for help in identifying the features. Looking at the quality of the pictures—up to the best professional standard—it is remarkable that all were taken with equipment and a 12-inch reflecting telescope made by Commander Hatfield himself.

Among popular accounts, V. A. Firsoff's *The Old Moon and the New* (Sidgwick & Jackson) and Thomas Rackham's *Moon in Focus* (Pergamon Press) are very good, though now slightly out of date. Among technical works there is of course R. B. Baldwin's *The Measure of the Moon* (Chicago University Press) which is devoted mainly to the impact theory of crater formation. As I was only one of the authors, I may perhaps be allowed to mention *The Craters of the Moon* by Peter Cattermole and myself (Lutterworth Press) which gives the opposite viewpoint.

I have several times mentioned the *Handbook* of the Lunar Section of the British Astronomical Association, which is devoted to the needs of the observer. It should not be confused with the Association's main *Handbook*; the official title of the Lunar Section booklet is *Guide for Observers of the Moon*.

There are two outline maps in general use. One was originally drawn in 1896 by T. Gwyn Elger, first Director of the B.A.A. Lunar Section, and is still available (George Philip & Co.), though unfortunately the new edition has followed the system of putting north at the top—making it inconvenient for the practical observer. The other map is my own, to a scale of 24 inches to the Moon's diameter.

Of the books dealing with Apollo results, I would make special mention of *Lunar Science: A Post-Apollo View*, by S. R. Taylor (Pergamon Press).

Appendix III

NUMERICAL DATA

Distance from Earth:

Mean,	238,840 miles or 0·0025695 astronomical units.
Maximum,	252,700 miles.
Minimum,	221,460 miles.

Sidereal period: 27·321661 days.

Synodic period: 29d 12h 44m 02s·9.

Axial inclination of equator, referred to the ecliptic: 1°32'.

Orbital eccentricity: 0·0549.

Orbital inclination: 5°09'.

Mean orbital velocity: 2,287 m.p.h.=0·63 miles per second =3,350 feet per second.

Apparent diameter: max. 33'31", mean 31'5", min. 29'22".

Magnitude of full moon, at mean distance: −12·7.

Mean albedo: 0·07.

Diameter: 2,160 miles (3,476 kilometres).

Mass: 1/81·3 Earth=0·0123 Earth=3·7 × 10^{-8} Sun.

Volume: 0·0203 Earth.

Escape velocity: 1·5 miles per second (2·38 km/sec).

Density: 3·34 water=0·60 Earth.

Surface gravity: 0·01653 Earth.

Appendix IV

ECLIPSES OF THE MOON, 1976–1986

Date.	Time of mid-eclipse, GMT. h m	Type.	Magnitude.	Duration of Totality. (minutes)	Moon overhead at: Long.	Lat.	Visible from: England and South Africa	U.S.A.
1976 13 May	19 50	Partial	0·14		66 E	18 S	Partly	No
1977 4 Apr.	04 21	Partial	0·21		64 W	06 S	Partly	Yes
1978 24 Mar.	16 25	Total	1·46	90	115 E	02 S	No	No
1978 16 Sept.	19 03	Total	1·33	82	73 E	03 N	Partly	No
1979 13 Mar.	21 10	Partial	0·89		44 E	03 N	Yes	No
1979 6 Sept.	10 54	Total	1·12	52	164 W	07 S	No	Partly
1981 17 July	04 48	Partial	0·58		71 W	21 S	Partly	Yes
1982 9 Jan.	19 56	Total	1·35	84	63 E	22 N	Yes	No
1982 6 July	07 30	Total	1·72	102	11 W	23 S	No	Yes
1982 30 Dec.	11 26	Total	1·20	66	171 W	23 N	No	Yes
1983 25 June	08 25	Partial	0·34		126 W	23 S	No	Yes
1985 4 May	19 57	Total	1·23	70	60 W	16 S	Partly	No
1985 28 Oct.	17 43	Total	1·08	42	90 E	13 N	Partly	No
1986 24 Apr.	12 44	Total	1·22	68	168 E	13 S	No	Partly
1986 17 Oct.	19 19	Total	1·27	74	67 E	10 N	Yes	No

The remaining total eclipses to the end of the century are those of 1989 20 February and 17 August, 1990 9 February, 1992 10 December, 1993 4 June and 29 November, 1996 4 April and 27 September, 1997 16 September, and 2000 21 January and 16 July.

The first four columns need no explanation. The fifth, headed Magnitude, is the extent of the eclipse, 1·0 or greater being total, anything less than 1·0 partial; for instance, 0·58 means that 58 per cent. of the Moon is covered by the umbra at mid-eclipse. Column 6 gives the geographical longitude and latitude on the Earth where the Moon is overhead at mid-eclipse. Columns 7 and 8 show whether the eclipse can be seen from England (and South Africa) or the United States. 'Partly' may mean that the eclipsed Moon is very low in the sky, or that the Moon rises or sets while the eclipse is in progress.

Appendix V

LUNAR LANDINGS

(a) Pre-Apollo automatic landing probes

Name	Launch date	Site	Notes.
Ranger 7	1964 28 July	Mare Nubium ('Mare Cognitum')	Crash-landed. Photographic.
Ranger 8	1965 17 Feb.	Mare Tranquillitatis	"
Ranger 9	1965 21 Mar.	Alphonsus	"
Luna 9	1966 31 Jan.	W. Oceanus Procellarum	Soft-lander. Photographic.
Surveyor 1	1966 30 May	N. of Flamsteed	"
Luna 13	1966 21 Dec.	W. Oceanus Procellarum	"
Surveyor 3	1967 17 Apr.	Oceanus Procellarum (Apollo 12 site)	Photographic; soil physics
Surveyor 5	1967 8 Sept.	Mare Tranquillitatis (15½ miles from Apollo 11 site)	Photographic: soil physics: chemical analyses
Surveyor 6	1967 7 Nov.	Sinus Medii	"
Surveyor 7	1968 17 Jan.	N. rim of Tycho	"
Luna 15	1969 13 July	Mare Crisium	Success only partial at best

(b) Russian Recoverable Missions

Name	Date	Site
Luna 16	Sept. 1970	Mare Fœcunditatis: lat. 0° 41' S, long. 56° 18' E
Luna 20	Feb. 1972	Apollonius highlands: lat. 3° 32' N, long. 56° 33' E

(c) *Russian 'Crawlers'*

Lunokhod 1	Nov. 1970	Western M. Imbrium: traverse length 12½ miles	
Lunokhod 2	Jan. 1973	Le Monnier, 110 miles N. of Apollo 17 site	" " 19 miles

(d) *Manned landings*

Mission	Date	Crew	Site	EVA dura- hours	Distance covered, miles
Apollo 11	1969 20 July	Armstrong, Aldrin, Collins	Mare Tranquillitatis, lat. 00° 67′ N, long. 23° 49′ E	2·2	—
Apollo 12	1969 19 Nov.	Conrad, Bean, Gordon	Oceanus Procellarum, lat. 03° 12′ S, long. 23° 23′ W	7·6	0·8
Apollo 14	1971 31 Jan.	Shepard, Mitchell, Roosa	Fra Mauro, lat. 03° 40′ S, long. 17° 28′ E	9·2	2·1
Apollo 15	1971 30 July	Scott, Irwin, Worden	Hadley-Apennines, lat. 26° 06′ N, long. 03° 39′ E	18·3	17
Apollo 16	1972 21 Apr.	Young, Duke, Mattingly	Descartes, lat. 08° 60′ S, long. 15° 31′ E	20·1	16
Apollo 17	1972 11 Dec.	Cernan, Schmitt, Evans	Taurus-Littrow, lat. 20° 10′ N, long. 30° 46′ E	22	18

Appendix VI

DESCRIPTION OF THE SURFACE, AND MAP

SINCE THIS BOOK is intended for observers, it seems only right to include a sectional map, but there are several points which I must make in connection with it. First and foremost, it does not set out to be a precision chart. It is no more than an outline, and is intended to do no more than act as a guide for those people who are trying to find their way about—and want to know which formation is which. I have included all the important names on the Earth-turned hemisphere, and the IAU directives have been followed throughout except that I have kept to some of the obvious English names for mountain ranges and rills. When the observer has identified all the formations given here, it will be high time for him to change to a larger-scale and more accurate map.

There are sixteen sections; 1 to 4 include the First Quadrant, 5 to 8 the Second, 9 to 12 the Third, and 13 to 16 the Fourth. There are bound to be some awkward subdivisions; for instance Archimedes, in the great Mare Imbrium group, lies in Section 5, while its companions Aristillus and Autolycus are in Section 4—it is a pity that the boundary had to come just there! However, I have linked each separate map with the sections which adjoin it. The descriptions are very short, and have had to be compressed almost to the point of dehydration, but again I hope that they will serve as a general guide. The depth and diameter measurements are approximate only, but I cannot think that it really matters whether the diameter of—say— Ptolemæus is 92 miles or 93; naturally I have given the figures as accurately as I can.

The map is drawn to mean libration, so that formations such as the Mare Orientale do not appear at all; but I have at least included notes about the few interesting features which are quite invisible except when the libration is at its best for that particular limb.

Finally: though the maps are drawn with south at the top,

I have followed the IAU convention with regard to east and west—so that anyone who happens to have an earlier edition of this book will find everything reversed. To recapitulate: Mare Crisium to the *east*, Grimaldi and Riccioli to the *west*.

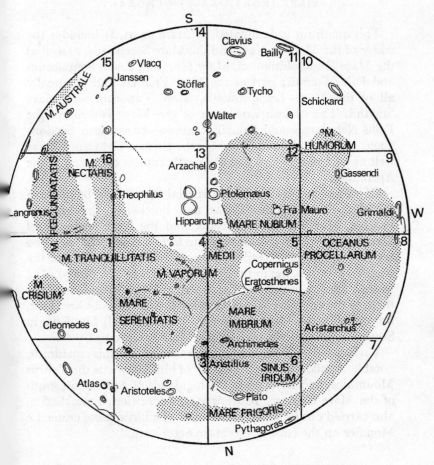

Key Map, showing sectional numbers in top right-hand corners

H*

This quadrant is dominated by Mare areas. It includes the whole of the Mare Crisium and the Mare Serenitatis, as well as the Mare Humboldtianum, Mare Marginis, Lacus Somniorum and Palus Somnii; part of the Mare Frigoris, and practically all of the Mare Tranquillitatis, Mare Vaporum and Mare Smythii. The eastern extensions of the Mare Imbrium—the Palus Nebularum and Palus Putredinis—extend into Sections 3 and 4. Mountain ranges include the Hæmus and Caucasus, as well as part of the Alps and the northernmost extension of the Apennines.

Of craters and walled plains, special note should be made of the dark-floored Julius Cæsar and Boscovich, the brilliant Proclus, Manilius and Menelaus, and the large enclosure of Cleomedes. Aristoteles and Eudoxus form a noble pair; here too are Autolycus and Aristillus, though the third member of the trio, Archimedes, lies in the Second Quadrant (Section 5). The Mare Vaporum is rich in rills; as well as Hyginus and Ariadæus there is the complex system associated with Triesnecker. Last, but certainly not least, the famous (or notorious!) Linné lies in Section 4, on the grey plain of the Mare Serenitatis.

Various space-craft have come down in this quadrant, notably Apollo 17 near the clump of hills known as the Taurus Mountains, and Luna 20 in the highlands of Apollonius, south of the Mare Crisium. Russia's second 'crawler', Lunokhod 2, also carried out its surveys here, near the incomplete crater Le Monnier on the edge of the Mare Serenitatis.

Section 1

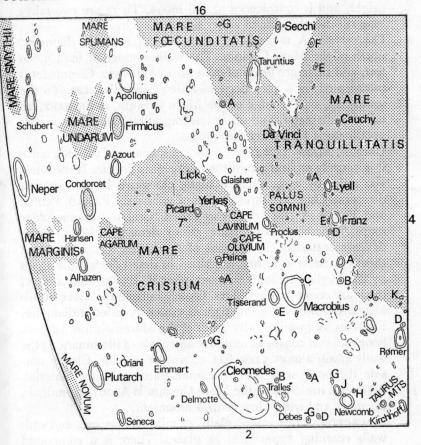

ALHAZEN. A crater 20 miles in diameter, near the border of the Mare Crisium. Roughly south of it lies a similar though slightly larger formation, HANSEN.

APOLLONIUS. A crater 30 miles in diameter, with walls rising to 5,000 feet above the floor. It lies in the uplands south of the Mare Crisium, and south again is the dark patchy area which has been called the 'Mare Spumans' or Foaming Sea on some maps, though it is certainly not a true Mare.

CAUCHY. An 8-mile crater on the Mare Tranquillitatis; it is bright, and is conspicuous at full moon. There are two fairly long rills nearby, one to either side of Cauchy.

CLEOMEDES. A magnificent enclosure 78 miles in diameter, north of the Mare Crisium. The walls average at least 9,000 feet, and there are peaks of even greater altitude. Cleomedes is interrupted by a very deep 28-mile crater, TRALLES. To the south-east, on the edge of the Mare Crisium, is EIMMART, 26 miles in diameter, and other less important formations in this area include DELMOTTE and DEBES.

CRISIUM, MARE. One of the most conspicuous of the seas, since it is entirely separate from the main Mare system. It measures 280 miles by 350, and is actually elongated in an east-west direction. On it are three craters of some size: PICARD, PEIRCE and PEIRCE A (the latter was called 'Graham' on the older maps). There are also many minor details, including some delicate craterlets closely west of the jutting CAPE AGARUM; these craterlets were first described in detail during the 1930s by the English amateur selenographer R. Barker, and some of them are connected by ridges. Closely outside the Mare in this area is CONDORCET, a fine regular crater 45 miles in diameter.

FIRMICUS. A 35-mile crater south of the Mare Crisium. Its dark floor makes it conspicuous under any angle of illumination; the walls attain almost 5,000 feet above the interior. Closely outside the north-west wall is a small patch of Mare material. Between Firmicus and the Mare Crisium is AZOUT, 19 miles in diameter and with a low central mountain.

MACROBIUS. A fine walled plain, 42 miles in diameter and with walls reaching 13,000 feet in places. There is a compound central mountain mass of moderate height. Between Macrobius and the Mare Crisium is a smaller crater, TISSERAND.

MARGINIS, MARE. A limb-sea; it is never well seen from Earth, but Orbiter and Apollo pictures have shown it to be a true Mare of the Crisium type, though less well-defined.

NEPER. A deep crater, 70 miles in diameter, between the Mare Marginis and the Mare Smythii. It is, of course, very fore-shortened as seen from Earth.

NEWCOMB. A 32-mile crater of considerable depth, west of Cleomedes. Its south wall is interrupted by a craterlet. New-comb is the northernmost and largest member of a string of three formations. East of it is a smaller but well-formed crater, KIRCHHOFF.

PICARD. The largest crater on the Mare Crisium; it is 21 miles in diameter, with walls rising to 8,000 feet. There is a central hill. To the west, on the edge of the Mare, lie the two capes LAVINIUM and OLIVIUM, together with imperfect craters such as LICK and YERKES.

PEIRCE. The second largest crater on the Mare Crisium; it is 12 miles in diameter, and the walls attain about 7,000 feet. North of it is the smaller but probably equally deep PEIRCE A.

PLUTARCH. A very foreshortened 40-mile crater with a central mountain. On the limb, to the north-east, is a small dark area which has been called the 'Mare Novum' or New Sea. Near Plutarch lies the rather smaller SENECA, while between Plutarch and Eimmart is the very ill-defined enclosure ORIANI.

PROCLUS. A brilliant crater west of the Mare Crisium, 18 miles in diameter and 8,000 feet deep. Proclus is one of the brightest points on the Moon, and is the centre of a ray-system; there is a low central mountain. The rays cross the Mare Crisium, but not the Palus Somnii, which is bounded by rays to either side. There are both bright and dusky bands on the inner walls of Proclus.

RØMER. A fine crater, 35 miles in diameter, with high terraced walls rising to over 11,000 feet. The floor contains a particularly large and massive central mountain, on the top of which is a summit pit. From Rømer the so-called TAURUS MOUNTAINS extend north-east, toward Newcomb and the well-marked Geminus (Section 2), but must be regarded as a hilly upland rather than a true range.

SMYTHII, MARE. Yet another limb-sea of the Crisium type, only well seen from space-probes. A good guide to it is the 46-mile crater SCHUBERT, which is not difficult to identify.

SOMNII, PALUS. Really an extension of the Mare Tranquillitatis, bounded on the north-west and south-east by rays from Proclus. The colour is peculiar, and has been described as brownish, greenish or yellowish; it is in any case different from that of the Mare Tranquillitatis itself. On the boundary between the Palus Somnii and the Mare Tranquillitatis are some low-walled craters, notably FRANZ and LYELL.

TARUNTIUS. An interesting crater, 38 miles in diameter, with low narrow walls nowhere attaining more than 3,500 feet. There is a central mountain crowned by a pit, and there is a complete inner ring on the floor, so that Taruntius is an excellent example of a 'concentric crater'. Well to the south-west lies SECCHI, an imperfect but quite conspicuous formation.

TRANQUILLITATIS, MARE. This is one of the major seas, and spreads from Section 1 into Section 4; it joins on to the Mare Serenitatis, Mare Vaporum, Mare Nectaris and Mare Fœcunditatis. The floor is decidedly lighter and patchier than that of the Mare Serenitatis, and the form is less regular. Important craters on it include Cauchy (Section 1) and Maskelyne and Arago (Section 4). It was, of course, in this Mare that Armstrong and Aldrin made their never-to-be-forgotten landing from the lunar module of Apollo 11.

UNDARUM, MARE. The so-called 'Sea of Waves' is a dark, patchy area, but not a well-defined sea; it lies near the Mare Crisium, east of Firmicus and Apollonius. It is very dark near full moon, and can be found without difficulty.

Section 2

ATLAS. A magnificent enclosure 55 miles in diameter. The much-terraced walls rise to as much as 11,000 feet above an interior which contains considerable detail—several old rings, some delicate rills, craterlets, and dark patches which seem to vary over the course of a lunation. These changes are, of course, due only to the altering angle of illumination; all the same, they are worth studying. Some way south-east of Atlas is the fine crater FRANKLIN, 34 miles in diameter, with walls reaching 8,000 feet; and in the Franklin area are various lesser en-closures such as CEPHEUS, ŒRSTED, CHEVALLIER and SHUCK-BURGH. Well south of Atlas is the bright, deep crater MAURY, 11 miles in diameter. The most important neighbour of Atlas is Hercules, with which it forms a noble pair; Hercules is described in Section 3.

BERNOUILLI. A 25-mile crater east of the larger Geminus. The walls of Bernouilli are highest on the east, where they reach up to about 13,000 feet above the floor. To the east is another crater-pair made up of BEROSUS and HAHN. Berosus, the larger

Section 2

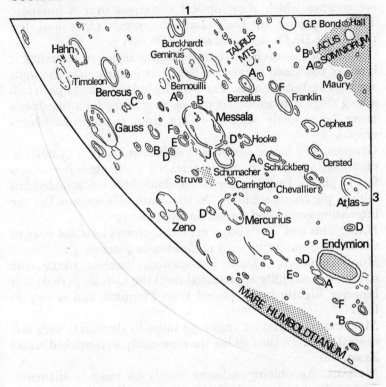

and more conspicuous, is 47 miles in diameter, with terraced walls.

BERZELIUS. A crater 24 miles in diameter. The floor is rather dark, and includes a central peak; the crater closely east also has a central peak.

CARRINGTON. An unremarkable crater between Messala and Mercurius.

CEPHEUS. The companion to Franklin; it is 28 miles in diameter, and its north-east wall is broken by a bright crater, Cepheus A. Between Cepheus and Franklin there are traces of an old ring.

ENDYMION. An important and interesting crater 78 miles in diameter. Its walls are high, containing a few peaks rising to 15,000 feet, and the floor is very dark, so that Endymion is easy to find under all conditions of lighting. Here, as in Atlas, there are patches which show optical variations over a lunation. The companion to Endymion, the much older and less prominent De La Rue, is shown in Section 3.

GAUSS. Since Gauss has a diameter in the region of 100 miles, it is one of the Moon's grand craters—particularly since the walls are high and continuous. The floor includes a central peak and a long ridge. Unfortunately it is so close to the limb that it can never be properly studied except from space-probe photographs.

GEMINUS. The larger companion of Bernouilli. It is 55 miles in diameter, with broad, richly terraced walls attaining 16,000 feet in places. In the centre of the floor there is a rounded hill with a pit on its summit. To the south of Geminus lies the interesting compound formation BURCKHARDT.

HALL. This and G. P. BOND are small craters near the edge of the Lacus Somniorum, east of Posidonius (Section 4).

HUMBOLDTIANUM, MARE. Yet another limb-sea whose true nature was not fully appreciated until the Orbiter period; it is however slightly better placed than Marginis, and is easy to find.

MERCURIUS. A distinct crater 33 miles in diameter, with low central peak. To the east lies the reasonably well-marked crater ZENO.

MESSALA. An oblong enclosure nearly 80 miles in diameter, with walls which are broken and generally low. Near it is HOOKE, only 27 miles across, but deeper and more distinct. Between Messala and Zeno are SCHUMACHER, smooth-floored and 25 miles across, and STRUVE, a small ring with a central peak, easy to recognize because it lies on a dark patch.

TAURUS MOUNTAINS. This upland area has been described in Section 1. It extends from the Berzelius area in the general direction of Rømer.

TIMOLEON. A large ring, 80 miles in diameter, southward along the limb from Gauss.

Section 3

ALPS. A bright mountain range forming part of the rampart of the Mare Imbrium, from Plato toward Cassini. Part of it is

Section 3

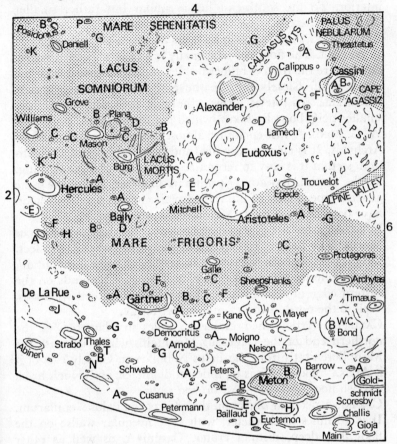

shown here, and part in Section 6. The peaks are moderately high; MONT BLANC (Section 6), near the great Valley, rises to 11,800 feet. Note too the small but rather bright crater TROUVELOT, 6 miles in diameter, right in the uplands. The ALPINE VALLEY is 80 miles long, and has been described in the text; it is by far the finest formation of its type on the entire Moon. The rill running down it is a very delicate object, though of course Orbiter and Apollo photographs show it well. ARCHYTAS. A bright crater 21 miles in diameter, on the north border of the Mare Frigoris. It has a triple-peaked central

241

mountain, and walls which rise to 5,000 feet above the sunken interior. To the south-east lies a similar but rather smaller crater, PROTAGORAS.

ARISTOTELES. A great plain 60 miles across, with walls rising to 11,000 feet. The floor contains many low hills, and particularly notable are the rows of hillocks which radiate outward from the crater itself. Closely outside the east wall is a deep crater, MITCHELL. Aristoteles forms a notable pair with its slightly smaller neighbour Eudoxus, which lies to the south.

W. C. BOND. A vast enclosure almost 100 miles across, north of Archytas. It is clearly very old, and is in a delapidated condition, so that it is easy to recognize only under oblique lighting. To the north is BARROW, 54 miles in diameter, and of the same general type. Many of the walled plains in this neighbourhood have been badly ruined; Meton is another example.

BÜRG. A crater 28 miles in diameter, near the edge of the Lacus Mortis. The floor is concave, and the walls rise to 6,000 feet. Bürg is notable both because of its very large central mountain, which includes a summit crater, and because it stands on the eastern edge of a small dark plain which is riddled with rills. Well to the north is the 12-mile crater BAILY.

CALIPPUS. A deformed crater 19 miles in diameter, at the northern end of the Caucasus Mountains which separate the Mare Serenitatis from the Palus Nebularum. Closely east of it is ALEXANDER, which is 65 miles in diameter, and which has a darkish floor and very low, broken walls.

CASSINI. A peculiar object on the edge of the Palus Nebularum. It is 36 miles in diameter, with very irregular walls; on the floor is a deep, distinct crater, Cassini A, as well as other details. The whole formation was strangely omitted from early maps, but there is not the slightest chance that it is of recent origin!

CAUCASUS MOUNTAINS. An important range, forming part of the border between the Mare Serenitatis and the Mare Imbrium (of which the Palus Nebularum is a part). Some of the peaks rise to 12,000 feet. The range extends from this Section into Section 4.

CHALLIS. Challis is 35 miles in diameter. It and its companion, MAIN, form a good example of overlapping; Main, 30 miles in diameter, is the intruding formation. The best 'pointer' in this area is the deep, distinct Scoresby. Challis and Main lie not far from the Moon's north pole, so that they are highly fore-shortened and are never well placed for observation from Earth.

DE LA RUE. A large enclosure not far from Endymion, and of about the same size. The walls are, however, very low and broken, and the formation is not distinct. Nearby are two much more prominent craters; Thales, which is a ray-centre, and Strabo.

DEMOCRITUS. A very deep ring 23 miles in diameter, of some-what distorted shape. It lies in the highlands north of the Mare Frigoris, close to the bay of Gärtner. There is a central mountain. To the west are two low-walled formations, KANE and MOIGNO. Moigno has a darkish floor. To the north-west of Democritus is the rather low-walled and obscure ARNOLD, 50 miles across.

EGEDE. A peculiar object shaped rather like a diamond; its mean diameter is 23 miles. The floor is dark, and the walls are very low. Egede lies roughly between Aristoteles and the Alpine Valley.

EUDOXUS. In many ways Eudoxus is similar to its companion Aristoteles, but it is smaller (diameter 40 miles) and lacks the remarkable radiating rows of hillocks. The walls attain 11,000 feet. Close to Eudoxus is the incomplete formation LAMÈCH.

FRIGORIS, MARE. This is an irregular Mare, and possibly in the nature of an overflow. Its floor is comparatively light and patchy. Craters on it include Galle and Protagoras. The Mare separates the Alpine region from the highlands around the north pole.

GALLE. A small but distinct crater on the Mare Frigoris, north of Aristoteles.

GARTNER. A splendid example of a bay. It lies on the border of the Mare Frigoris, and the Mare-material has reduced its 'seaward' wall to such an extent that it is now barely traceable, though the 'landward' rampart is still quite high. Gärtner is 63 miles in diameter, and the floor contains some delicate rills.

GIOJA. A 26-mile crater close to the north pole, and thus very badly placed for observation. It abuts into a larger formation with low walls.

GOLDSCHMIDT. An old ring, 68 miles in diameter, between

Barrow and the very prominent Anaxagoras in the Second Quadrant (Section 6). Its walls are low and broken.

HERCULES. The smaller companion of Atlas. Hercules is 45 miles across, and has walls rising to 11,000 feet; these walls are richly terraced, and often appear brilliant. The floor contains one prominent crater as well as a large amount of fine detail. Several TLPs have been reported here over the years. Well south of Hercules is WILLIAMS, just on the Lacus Somniorum.

MASON. This crater is 15 miles in diameter; it forms a pair with Plana (24 miles) not far from Bürg. Both craters have low, broken walls. Plana has the darker floor. South-east of Mason, on the border of the Lacus Somniorum, is a deeper and more distinct crater, GROVE.

MORTIS, LACUS. A small darkish plain near Bürg.

METON. Another of the large enclosures in the north polar region. It lies not far from the distinct Scoresby. Meton is over 100 miles in diameter, but is really a compound formation, made up of several ringed plains which have run together. Between it and the limb, close to its wall, is a smaller but more perfect enclosure, EUCTEMON, and to the north-east is the broken crater BAILLAUD. Near the limb in this area are various other craters of some size, notably PETERMANN and CUSANUS, all of which are naturally very foreshortened. Smaller rings include PETERS, C. MAYER and NEISON.

NEBULARUM, PALUS. Part of the Mare Imbrium; the region of Aristillus and Autolycus (Section 6). The name is an old one, but has been omitted on some modern official maps.

PLANA. The companion of Mason, and described with that crater.

SCHWABE. A small, deep crater north of Gärtner, in the highlands.

SCORESBY. A very deep, distinct crater 36 miles in diameter, in the north polar uplands near Challis. It has a twin-peaked central hill. Scoresby is usually easy to recognize, since it is much better-formed than any of its immediate neighbours.

SHEEPSHANKS. A crater just on the Mare Frigoris, well northwest of Aristoteles.

SOMNIORUM, LACUS. This is really a bay leading out of the Mare Serenitatis, but its floor is much lighter and patchier. On its southern border is the great enclosure Posidonius (Section 4), but the 18-mile DANIELL, the well-formed companion of Posidonius, is in this Section. There is some fine detail on the Lacus, including some delicate rills.

STRABO. A 32-mile crater close to De La Rue. The wall contains some high peaks, and the floor is comparatively smooth. Strabo is the centre of a short and inconspicuous ray system, much less prominent than that of its neighbour Thales.

THALES. Thales, close to Strabo, is 24 miles across, and is a major ray-centre, so that it is very prominent near full moon.

THEÆTETUS. A crater on the Palus Nebularum, near Cassini; it is 16 miles in diameter, and has a low central mountain. It was near Theætetus that Charbonneaux recorded his 'cloud' in 1902, described in the text.

TIMÆUS. A bright crater, 21 miles in diameter, on the north border of the Mare Frigoris, not far from Archytas. The floor contains a double central hill. Timæus is the centre of a minor ray system, and acts as a good guide to the large, broken enclosure, W. C. Bond.

Section 4

AGRIPPA. A fine crater 30 miles in diameter not far from Hyginus, near the border of the Mare Vaporum. Its walls, which rise to 8,000 feet, are terraced; and there is a central mountain. It forms a notable pair with its slightly smaller southern neighbour, Godin.

APENNINES. The Apennines are certainly the most spectacular mountains on the Moon, though not the highest; they extend from this Section into Section 5, ending near Eratosthenes. They form part of the rampart of the Mare Imbrium, and make a magnificent spectacle when suitably lit. The loftiest peak in this Section is MOUNT BRADLEY, near Conon, which rises to 16,000 feet; MOUNT HADLEY, at the northern end of the range, is only a thousand feet lower than this, and it was in the foot-hills of the range in this area that the astronauts of Apollo 15 made their landing. The range ends, to the north, at the jutting CAPE FRESNEL, after which there is a gap between the Mare Imbrium and the Mare Serenitatis until the border is resumed with the Caucasus Mountains. There are various craters in the Apennine uplands, notably the 13-mile CONON.

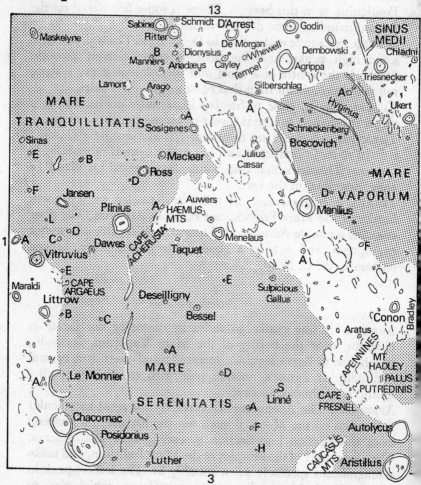

ARAGO. A distinct crater, 18 miles in diameter and obviously distorted from the circular form. It lies on the Mare Tranquillitatis; the floor includes a central elevation. A low, imperfect ring, LAMONT, lies to the south-east, and the bright 10-mile crater MANNERS to the south-west. Arago is notable because there are various domes to the west of it. These are among the best examples of domes on the whole of the Moon, and are worth studying; several contain summit pits.

ARATUS. A small bright crater in the Apennines, south of Cape Fresnel.

ARGÆUS, CAPE. A high promontory, guarding the strait between Mare Tranquillitatis and Mare Serenitatis.

ARIADÆUS. A bright 9-mile crater, with a smaller one in contact with it on the north-east. It lies on the border of the Mare Tranquillitatis, and is notable because of the great rill nearby, discovered by Schröter in 1792 and easy to see even with a small telescope. It is over 150 miles long, and has various branches, one of which connects the system with that of Hyginus. The main rill runs across the uplands into the Mare Vaporum.

ARISTILLUS. A splendid crater on the Palus Nebularum, forming a trio with Archimedes (Section 5) and Autolycus. Aristillus is 35 miles in diameter, with walls rising in places to 11,000 feet. The interior details show optical variations over a lunation. The floor includes a fine triple-peaked central mountain.

AUTOLYCUS. The southern companion of Aristillus. It is smaller (24 miles in diameter) but just as distinct, and the walls rise to 9,000 feet. Under high light Autolycus is seen to be the centre of an inconspicuous ray system, and under a low sun radiating ridges can be made out extending from it.

BESSEL. The largest crater on the Mare Serenitatis; it is 12 miles in diameter, with walls rising to 3,600 feet above the depressed interior. The great ray crossing the Mare Serenitatis passes nearby. To the east lies a smaller crater, DESEILLIGNY.

BOSCOVICH. A low-walled formation on the border of the Mare Vaporum. It is about 27 miles across, but is irregular in form. It is notable because of the darkness of its floor, which makes it very easy to recognize; the same is true of its neighbour Julius Cæsar.

CAYLEY. One of the bright craters in the uplands between the Mare Vaporum and the Mare Tranquillitatis. It is 9 miles in diameter, and very distinct. The nearby TEMPEL and DE MORGAN are also bright, but only about 5 miles in diameter; WHEWELL is of the same type.

CHACORNAC. A pentagonal ring-plain about 30 miles in diameter, close to Posidonius on the edge of the Mare Serenitatis. There is considerable detail on the floor, including one distinct crater, A.

DAWES. A 14-mile crater between the Mare Serenitatis and the

Mare Tranquillitatis; it is somewhat deformed. Two dusky bands run from the small central peak up the inner west wall.

DIONYSIUS. A brilliant crater, 12 miles in diameter, on the edge of the Mare Tranquillitatis not very far from Ariadæus. It stands on a light area, and is very conspicuous under high illumination. It is yet another crater with dark bands running up its walls.

GODIN. A crater 27 miles in diameter, slightly deformed, but with a central hill. It is the southern member of the Godin-Agrippa pair. Between it and the Sabine-Ritter pair lies D'ARREST. The low-walled DEMBOWSKI lies some way west of Godin and Agrippa.

HÆMUS MOUNTAINS. These mountains form part of the border of the Mare Serenitatis, and separate it from the Mare Vaporum. They are not lofty, but some of the peaks rise to about 8,000 feet. The glittering Menelaus lies on the edge; so does the much smaller and less bright AUWERS.

HYGINUS. A depression about 4 miles in diameter, notably because of its association with the famous crater-rill which has been fully described in the text. North of it lies Hyginus N, and here too is an interesting spiral mountain, MOUNT SCHNECKEN-BERG, which requires a fairly high power to be well seen. Some branches of the Hyginus rill-system join up with the system of Ariadæus.

JULIUS CÆSAR. An imperfect, very dark-floored enclosure between the Mare Tranquillitatis and the Mare Vaporum, not far from Boscovich. This is one of the darkest patches on the entire Moon. Outside its east wall is a crater-valley.

LE MONNIER. A fine example of a bay; it lies on the border of the Mare Serenitatis, and is 34 miles in diameter. Of the sea-ward wall, only a few mounds now remain; the floor contains little visible detail. It was in this region that Russia's Lunokhod 2 came down—and where it still remains!

LINNÉ. This celebrated formation, on the Mare Serenitatis, has been fully described in the text.

LITTROW. A 22-mile crater between Le Monnier and Cape Argæus, on the edge of the Mare Serenitatis. The walls are of some height, but are broken in places. Apollo 17 came down in the Taurus-Littrow area.

MANILIUS. A splendid crater 25 miles in diameter, on the

border of the Mare Vaporum. Its walls are brilliant, so that Manilius is very prominent near full moon. The walls are terraced, and there is considerable interior detail, including a central mountain.

MARALDI. A low-walled crater near Rømer (Section 1).

MASKELYNE. A 19-mile crater on the Mare Tranquillitatis, not very far from the Apollo 11 landing site. The walls have inner terraces, and there is a low central peak. To the west is a conspicuous little crater, Maskelyne B.

MENELAUS. A brilliant crater in the Hæmus Mountains, striking under a high light even though it is only 20 miles in diameter. The walls rise to 8,000 feet above a floor which contains a peak not quite centrally placed.

PLINIUS. A superb crater 'standing sentinel' on the strait between the Mare Serenitatis and the Mare Tranquillitatis. It is 30 miles across, but appreciably distorted from the circular form. The central structure takes the form of a twin crater. Plinius has high terraced walls, and is conspicuous under all conditions of illumination.

POSIDONIUS. A walled plain 62 miles in diameter, adjoining Chacornac and lying on the borders of the Mare Serenitatis and the Lacus Somniorum. The ramparts are rather low and narrow, and the floor is crowded with detail, including a nearly central craterlet, Posidonius A. On the Mare, to the west, is the craterlet LUTHER.

PUTREDINIS, PALUS. Part of the Mare Imbrium, north-west of Mount Hadley. It spreads on to Section 5.

RITTER. This and its neighbour Sabine form a striking pair on the Mare Tranquillitatis, again in the general area of Apollo 11. Ritter is very slightly the larger, and is about 19 miles across. Nearby is a small but bright crater, SCHMIDT, and closely north of Ritter are two more bright craterlets, making almost perfect twins of the sort so common on the Moon. Both Sabine and Ritter have central peaks.

ROSS. A crater with a central peak. It is 18 miles in diameter, on the Mare Tranquillitatis. To the north-east is the craterlet Ross D, where TLP have been reported—though in my view the evidence is very slender. To the south-west of Ross lies MACLEAR, rather dark-floored and 11 miles across, west of which is a fine long rill running to, and beyond, Sosigenes A.

SABINE. The companion of Ritter, and described with it.

SCHNECKENBERG, MOUNT. The strange spiral mountain, described with Hyginus.

SERENITATIS, MARE. One of the most perfect of the regular seas. It covers an area of 125,000 square miles, slightly more than that of Great Britain. It is bordered, in part, by the Caucasus Mountains and by the Hæmus Mountains, ending at CAPE ACHERUSIA. The floor is relatively smooth, but contains several craterlets of which the most prominent are Bessel, Deseilligny and Luther; of course Linné also lies on the Mare. There are many prominent ridges, some of which seem to be the walls of ghost-craters, and the controversial bright ray runs right across the Mare, from Menelaus through to Bessel, Luther and on into the highlands beyond.

SILBERSCHLAG. A small, bright crater 8 miles in diameter, near Ariadæus. It is not unlike Cayley, though slightly smaller.

SINAS. This and its smaller companion, E, lie on the Mare Tranquillitatis, well north of Maskelyne.

SOSIGENES. A 14-mile crater to the east of Julius Cæsar; it has a small central hill, and its walls are bright. To the south-east is Sosigenes A, which lies on a long rill running from Maclear, and which is connected to Sosigenes itself by a ridge.

SULPICIUS GALLUS. An extremely bright crater, 8 miles in diameter, and with walls which rise to about 8,000 feet. It lies just on the Mare Serenitatis, in the foothills of the Hæmus Mountains. There is a small central peak.

TAQUET (sometimes spelled 'Tacquet'). Another bright crater-let just on the Mare Serenitatis, in the Hæmus region; it is 6 miles across.

TRANQUILLITATIS, MARE. A major sea, extending on to Sections 1 and 13. It is less regular in outline than the neigh-bouring Mare Serenitatis, and has a lighter, patchier floor. On it are various craters; some have been already been described, and others include the regular SINAS and the low-walled JANSEN.

TRIESNECKER. A crater in the area of the Mare Vaporum and of the Sinus Medii, which is described with Section 12; west of it is the distinct crater CHLADNI. Triesnecker is 14 miles across, and is notable because of the very complex rill-system to the east of it. The chief rills are visible with a small telescope under good conditions.

UKERT. Another 14-mile crater at the edge of the Mare Vaporum, north of Triesnecker. Here too there are rills, though the system is a minor one and does not rival that of Triesnecker. Ukert itself has rather bright walls.

VAPORUM, MARE. One of the minor seas, but notable because of its darkness and because of the many interesting objects nearby—such as Boscovich, Manilius and the rill-systems of Ariadæus, Hyginus, Triesnecker and Ukert. A small part of it extends on to Section 5.

VITRUVIUS. An interesting formation, just on the Mare Tranquillitatis near Mount Argæus. The walls are bright, but the floor is decidedly dark, containing a low central peak. Well to the south-west of it, right on the Mare, is Jansen, which is 16 miles across and has very low walls, rising to only about 300 feet above its darkish floor—which is scarcely depressed below the general level of the Mare. About the same distance southwest of Jansen is the small, well-marked crater Sinas. Northeast of Vitruvius, in the uplands, lies MARALDI, which is reasonably well-formed.

SECOND (NORTH-WEST) QUADRANT

This is the 'sea' quadrant, and there is little true upland. Most of the area is covered by the two greatest maria on the Moon—the Mare Imbrium and the Oceanus Procellarum, together with various minor seas such as the Sinus Æstuum, the Sinus Roris and the lovely Sinus Iridum or Bay of Rainbows. Only along the limb do we find tremendous walled plains, of which Pythagoras and Xenophanes are good examples. On the Mare surface lie three of the most famous of all lunar craters: Copernicus, Archimedes and Aristarchus, while the dark-floored Plato is to be found on the border of the Mare Imbrium.

The southern part of the Apennine range lies in this quadrant, ending near the majestic Eratosthenes; there are also the less lofty Carpathian Mountains, near Copernicus, and the Jura Mountains, bordering Sinus Iridum. There are also some lofty peaks near the limb.

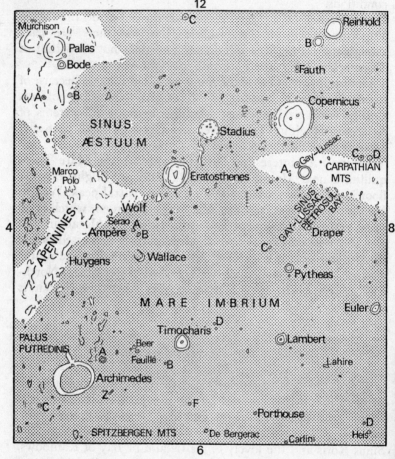

Section 5

ÆSTUUM, SINUS. A conspicuous dark plain east of Copernicus, bordered by the southernmost extension of the Apennines. Eratosthenes lies at its boundary. The floor is relatively devoid of important craters.

APENNINES. This superb range stretches up from Section 4, ending at Eratosthenes. This part of it contains the highest peak, MOUNT HUYGENS (18,000 feet); there is also the triangular mountain mass MOUNT WOLF, at least 12,000 feet, and MOUNT AMPÈRE, around 11,000 feet. There are a few low-walled, rather distorted craters in the Apennine uplands, such as MARCO POLO, 12 miles across with a darkish floor, and SERAO.

ARCHIMEDES. The largest of the craters on the Mare Imbrium; it forms a trio with Aristillus and Autolycus (Section 4). Archimedes is 50 miles in diameter, with walls around 4,300 feet above the slightly sunken floor; the rampart includes a few peaks reaching to at least 7,000 feet, but in general the wall has been much reduced. The floor is dark and (by lunar standards) very smooth, with no vestige of a central mountain.

BEER. An 8-mile crater on the Mare Imbrium, between Archimedes and Timocharis. It has an almost identical twin, FEUILLÉ, closely north-west of it. Like those of Messier and Messier A, the relative sizes of Beer and Feuillé seem to vary; these apparent changes are due only to optical effects, but they are interesting nonetheless. Between Archimedes and the Beer-Feuillé pair is a small deep crater, A, with a central peak.

BODE. An 11-mile bright crater in the highlands separating the Sinus Medii from the Sinus Æstuum. Its walls reach to 5,000 feet above the floor. Bode lies close to the semi-ruined crater-ring Pallas, and is conspicuous near full moon; it is the centre of a minor ray system.

CARLINI. A small, rather bright crater on the Mare Imbrium. It is 5 miles in diameter, with a small central hill; its isolated position makes it conspicuous.

CARPATHIAN MOUNTAINS. A range forming part of the border of the Mare Imbrium, in the Copernicus area. It cannot compare with the Apennines, and is rather discontinuous, though it extends for a total of between 200 and 250 miles. Its highest peaks attain 7,000 feet, but most of the mountains are much lower than this. The western end of the range stretches into Section 8.

COPERNICUS. The 'Monarch of the Moon'. It has been described in the text, and no more need be said here except to repeat that it and its rays dominate this whole part of the lunar surface.

ERATOSTHENES. This also ranks as one of the most perfect craters on the Moon. It is 38 miles in diameter, and extremely deep, with central elevations and much floor-detail. It marks the southern end of the Apennine range.

EULER. A minor ray-centre on the Mare Imbrium, and therefore easy to find under a high light. Euler is a well-marked crater 19 miles in diameter, with a central peak and walls which show some inner terracing.

FAUTH. A double craterlet south of Copernicus; its form makes it easy to identify. This whole area is dominated by ridges and crater-chains radiating from Copernicus. Fauth was splendidly shown on the celebrated 'Picture of the Century' of Copernicus and its environs, taken by Orbiter 2—the best lunar probe picture taken up to that time, though of course surpassed many times since.

GAY-LUSSAC. An irregular 15-mile crater in the Carpathian Mountains, north of Copernicus. Its floor contains some fine detail, including delicate rills. To the west, in the direction of Tobias Mayer, are two bays in the Carpathian range which have been named SINUS GAY-LUSSAC and PIETROSUL BAY. Immediately south-east of Gay-Lussac itself is the smaller, deeper crater Gay-Lussac A, and to the north of the bays are two more small craterlets, DRAPER and Draper C.

HEIS. A small craterlet on the Mare Imbrium, between De L'Isle (Section 8) and Caroline Herschel (Section 7). Heis is very minor, but is easy to find because of its isolated position.

IMBRIUM, MARE. The greatest of all the regular seas. Most of it lies in this Section, but it also extends into Sections 6 and 8. It has been fully described in the text, but it is worth repeating that in area it is larger than Great Britain and France combined.

LA HIRE. A bright 5,000-foot mountain on the Mare Imbrium, north-west of Lambert. It has a summit craterlet.

LAMBERT. A crater only 18 miles across, and not bright; but it is easy to find, owing to its position on the Mare Imbrium. In place of a central mountain, it has a central crater—a type of feature not uncommon on the Moon.

MURCHISON. The companion of Pallas, on the edge of the Sinus Medii. It has been badly distorted, and is clearly very old, so that its walls are now low and broken. It is about 35 miles in diameter.

PALLAS. Pallas, adjoining Murchison, is rather smaller (diameter 30 miles) but more complete, even though its walls have been broken by mountain passes. The central peak still exists. On the other side of Pallas is the smaller but much deeper crater Bode.

PUTREDINIS, PALUS. Part of the Mare Imbrium, to the west of Archimedes.

PYTHEAS. A very bright crater on the Mare Imbrium, with terraced walls and a central hill. It is a minor ray-centre, and is so conspicuous that it is surprising to find that its diameter is a mere 12 miles. Well to the south lie Draper and Draper C. Rays from Copernicus cross this whole area.

REINHOLD. A 30-mile crater on the Oceanus Procellarum, south-west of Copernicus; its walls rise to 9,000 feet in places. Closely north-east of it is Reinhold B, which has low walls and is about 15 miles across, with a darkish floor.

STADIUS. The celebrated 'ghost', on the edge of the Sinus Æstuum. It forms a triangle with Copernicus and Eratosthenes. It has been described in the text.

TIMOCHARIS. An interesting crater on the Mare Imbrium, roughly west of Archimedes. It is 25 miles in diameter, with broad terraced walls reaching to 7,000 feet. Like Lambert, it has a central crater. Timocharis is the centre of a rather faint ray-system, and is easy to identify.

WALLACE. An incomplete crater between Archimedes and Eratosthenes.

Section 6

ANAXAGORAS. A crater 32 miles in diameter, in the north polar uplands and closely west of Goldschmidt (Section 3), which it distorts. Anaxagoras is well-formed, with high walls and a central mountain. It is extremely bright, and is the centre of a major ray-system, so that it is easy to find under all conditions of illumination. It is a pity that it is not better placed for observation from Earth.

ANAXIMANDER. A walled plain 54 miles in diameter, with a good deal of floor-detail (though no central peak) and walls which rise in places to 9,000 feet. Adjoining it to the north-east is the smaller but almost equally deep CARPENTER. The limb-formations in this area are of great interest, and there are some splendid walled plains which come into view only under

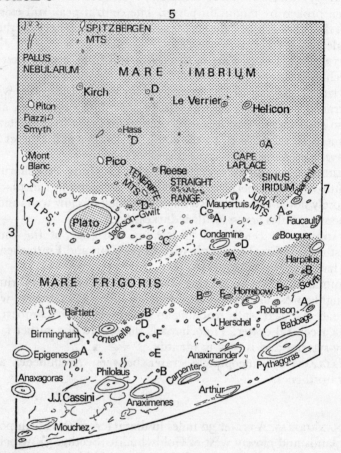

conditions of extreme libration, so that they cannot be shown on these maps. The newcomer to lunar work is often surprised to see them—as I originally was!

BIANCHINI. A bright-walled, somewhat irregular crater in the Jura Mountains, bordering Sinus Iridum. It has a central peak, and is usually easy to find. Its diameter is about 25 miles. To the east of it, also in the Jura range, is the very low-walled, irregular enclosure MAUPERTUIS.

BIRMINGHAM. A very large enclosure, about 66 miles across, in the highlands north of the Mare Frigoris, and east of Fontenelle.

It has been much distorted, and is crowded with detail, but it is not too easy to identify unless really well-placed with regard to illumination.

CONDAMINE. A 30-mile crater in the foothills of the Jura Mountains, on the border of the Mare Frigoris. Its walls are broken by passes and one distinct crater. To the north-west is a distinct, deep crater, Condamine A.

EPIGENES. A 30-mile crater with broad walls, south of Anaxagoras.

FONTENELLE. A bright walled plain with a central crater. It lies on the northern edge of the Mare Frigoris, and is easy to find on account of its brilliance. Close by it is the famous 'Mädler's Square', described in the text; and north of Fontenelle there is an ill-formed plain, J. J. CASSINI, bounded by irregular ridges.

FRIGORIS, MARE. The western part of the Mare Frigoris extends into this Section, from Section 3. The general aspect is the same as that of the eastern part.

HARPALUS. A conspicuous crater 32 miles across, near the borders of the Mare Frigoris and the Sinus Roris. It has an asymmetrical floor-mountain. Not far from it are two more distinct, rather deep craters, FOUCAULT and BOUGUER, both much smaller than Harpalus. Incidentally, Harpalus was the crater selected as a rocket launching-site in the famous film *Destination Moon*, produced in the early 1950s when space-travel was still regarded as something of a joke!

HELICON. This and Le Verrier form a conspicuous pair on the Mare Imbrium, near the Sinus Iridum. Helicon is 13 miles in diameter, Le Verrier 11; Helicon has a central craterlet, Le Verrier a central peak. Both have moderately high walls. Oddly enough Helicon is always easy to find, but Le Verrier becomes so obscure near full moon that it is difficult to locate at all.

HERSCHEL, JOHN. A ridge-bordered enclosure on the edge of the uplands bordering the Mare Frigoris to the North. It is about 90 miles across, and its floor is very rough. To the south-west are several distinct craters, including HORREBOW and ROBINSON.

IMBRIUM, MARE. Part of the vast Mare Imbrium extends into this Section. The general aspect is the same as that of the area in Section 5.

IRIDUM, SINUS. The lovely 'Bay of Rainbows'; one of the most spectacular sights on the Moon near sunrise, when the floor is in shadow and the Jura Mountains are illuminated, so that they seem to protrude into the blackness and give the 'jewelled handle' effect. From the Mare, the level of the Bay gradually slopes down to the extent of 2,000 feet. The seaward wall has almost vanished; its site is now marked only by a few very low, irregular ridges and one or two small craterlets. The capes to either side of it are HERACLIDES and LAPLACE.

KIRCH. A bright 7-mile crater on the Mare Imbrium, north of the Spitzbergen Mountains. Well to the north-west lies a somewhat smaller crater-pair.

LE VERRIER. This is the companion of Helicon, and has been described with it.

NEBULARUM, PALUS. Part of the Mare Imbrium, extending into this Section from Section 3.

PHILOLAUS. A very deep, prominent crater 46 miles across, with walls rising to 12,000 feet above a floor which contains a good deal of detail. Unusual colour effects have been reported here. ANAXIMENES adjoins Philolaus to the west; it is slightly the larger of the two, but not nearly so deep or conspicuous. To the east of Philolaus is the ill-formed J. J. Cassini, already described.

PICO. A bright mountain on the Mare Imbrium, south of Plato. It is triple-peaked; the maximum height is about 8,000 feet. The area between Pico and Plato is occupied by a very large ghost-ring; it was once known as Newton, though the name has now been transferred to a crater in the opposite part of the Moon. The 'ghost' is still sometimes referred to as Ancient Newton. The bright TENERIFFE MOUNTAINS, some of which are almost as lofty as Pico, also lie near the boundary of the destroyed ring, and may once have formed part of its wall, though more probably they are of considerably later date.

PITON. Another isolated mountain, 7,000 feet high and with a summit craterlet. It lies well to the south-east of Pico, and is always easy to find, because of its brightness. Between the two peaks, rather closer to Piton, is the bright 6-mile crater PIAZZI SMYTH, and there are various other mountains and small craterlets in the area.

PLATO. Hevelius' 'Greater Black Lake'; the 60-mile crater

noted for the darkness of its floor, which makes it easy to recognize under any conditions of lighting. Many TLP have been reported here. To the north-west is the well-marked deep crater Plato A, named on some old maps 'Jackson-Gwilt'.

PYTHAGORAS. One of the grand craters of the Moon, with its lofty, continuous walls and central mountain mass. It is, unfortunately, too foreshortened to be seen properly from Earth, but space-probe photographs bring it out in its true guise.

SOUTH. This is a ridge-bounded enclosure about 60 miles across. It and its larger and even more ruined neighbour BABBAGE lie close to Pythagoras, and are full of detail, but their walls are so broken and discontinuous that they are often difficult to recognize. The nearby Robinson is only 17 miles in diameter, but is generally easier to locate.

SPITZBERGEN MOUNTAINS. A series of bright little hills north of Archimedes (Section 5). They lie on the edge of a very obscure ghost-ring which is now traceable only because of a slight difference in the hue of what must once have been its floor.

STRAIGHT RANGE. A remarkable line of peaks, rising to a maximum of 6,000 feet, on the Mare Imbrium between Plato and Cape Laplace. The peaks are less brilliant than Pico or Piton, but are still quite bright. The range is curiously regular, and there is nothing quite like it anywhere else on the Moon.

TENERIFFE MOUNTAINS. These little peaks lie near Pico, on the border of 'Ancient Newton', the ghost-ring between Pico and Plato. They have been described with Pico.

Section 7

CLEOSTRATUS. A large, well-formed enclosure very close to the limb, not far from Xenophanes; there are various other rings in the libration zone. Cleostratus has steep, rather narrow walls.

GERARD. Another large, well-formed limb feature, with a long ridge running down its floor. Further on the disk is a 14-mile crater, HARDING, with low walls; north-east of Harding is another small crater, DECHEN. Other limb-craters in this area are LA VOISIER and (not shown on a mean libration map) RÉGNAULT, GALVANI and various others. The crater NAUMANN, some way from La Voisier, has fairly bright walls.

HERSCHEL, CAROLINE. An 8-mile crater on the Mare Imbrium,

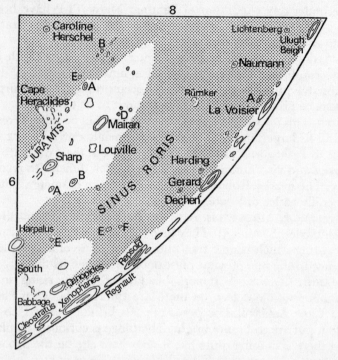

forming a triangle with Carlini (Section 5) and De L'Isle (Section 8). Its isolated position makes it conspicuous, particularly as its walls are rather bright.

IRIDUM, SINUS. A small part of the Sinus, including Cape Heraclides, appears in this Section, but most of it is in Section 6.

LICHTENBERG. A small crater on the Oceanus Procellarum. It is 12 miles in diameter, and is a minor ray-centre, though the rays are very short. Under high light, Lichtenberg appears as an ill-defined whitish patch. Reddish 'events' have been reported here now and then ever since the time of Beer and Mädler. Between Lichtenberg and the limb is a well-formed crater, ULUGH BEIGH, which is 30 miles across and has a central peak.

MAIRAN. A fine, well-formed crater in the Jura uplands. Its walls are lofty, but there seems to be no central mountain. North of it, also in the Jura uplands, is the ill-defined depression

LOUVILLE, which is not hard to identify because of its dusky floor.

ŒNOPIDES. A prominent crater near Cleostratus and Xenophanes. It is 42 miles in diameter, and has high walls, broken in the south-west by a craterlet. There are some minor features on the floor, though, like many of its companions in this part of the Moon, it lacks a central peak. Immediately east of it lies Babbage, which has already been described (Section 6), and to the south-west lies the bright, regular crater Œnopides A.

PROCELLARUM, OCEANUS. Part of this vast plain is shown here, extending from the south into Sinus Roris.

REPSOLD. Yet another enclosure very near the limb, west of Xenophanes.

RORIS, SINUS. This Bay forms the 'outlet' of the Oceanus Procellarum into the Mare Frigoris. It is not particularly remarkable, and its surface has the same general aspect as that of the Oceanus. On it are a few minor features, together with one peculiar formation, RÜMKER, which seems to be in the nature of a semi-ruined plateau about 30 miles across.

SHARP. A deep crater 22 miles in diameter, in the Jura Mountains and surrounded by high peaks. The floor includes a small central peak. Two small, fairly deep craters, A and B, lie to the north-west.

ULUGH BEIGH. This large crater lies near Lichtenberg, and has been described with it.

XENOPHANES. A grand 67-mile crater, with a massive, elongated central mountain crowned by a craterlet. The walls are lofty and terraced. However, it is very badly placed, and is much too near the limb to be well seen even when libration is at its maximum.

Section 8

ARISTARCHUS. The brightest formation on the Moon—and the most 'active', since gaseous emissions have been proved, and the area including Aristarchus, Herodotus, Prinz and the Harbinger Mountains has been responsible for more than half the number of TLP reported over the years. As early as 1911 R. W. Wood, in the United States, took some ultra-violet photographs which led him to believe that a small area near Aristarchus was

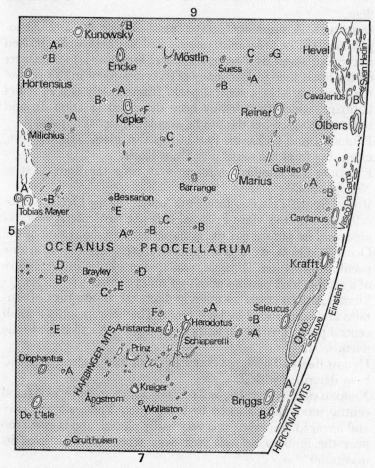

covered with a sulphur deposit, or at any rate something quite unlike the surrounding regions; this is still sometimes referred to as 'Wood's Spot'. The extreme brilliance of Aristarchus makes it obvious under any conditions, even when illuminated only by earthshine.

BESSARION. A bright 6-mile crater on the Oceanus Procellarum. It has a central hill, and a dark band up the inner south-west wall. North of Bessarion is a smaller but equally bright crater, Bessarion E, sometimes called VIRGIL.

BRAYLEY. Another crater on the Oceanus Procellarum, similar to Bessarion but rather larger (diameter 10 miles). It too has a low central hill and dusky radial bands in its interior. It is a member of a curved line of craterlets, of which B, south-east of Brayley itself, has rather bright walls.

BRIGGS. A 33-mile crater on the Oceanus Procellarum, east of Otto Struve. It is well-marked, and easy to locate. Ridges connect it with Seleucus.

CARDANUS. This and KRAFFT form another notable pair in the limb-region at the very edge of the Oceanus Procellarum. Cardanus has a diameter of 32 miles and continuous walls attaining 4,000 feet. It has a central mountain, and there are numerous craterlets on its floor. To the south-east, between Cardanus and Reiner, lies the bright little crater GALILEO (more properly, Galilaei), which is fairly bright—yet a 9-mile formation is surely inadequate to honour the first man to undertake serious, regular telescopic studies of the Moon! As has been noted in the text, Riccioli, who named the crater in his 1651 map, disliked Galileo as being a protagonist of the heretical theory that the Earth moves round the Sun instead of vice versa, so that when the names were being allotted Galileo was treated very badly indeed.

CAVALERIUS. The northern member of the chain which includes Hevel as well as Lohrmann, Riccioli and Grimaldi (Section 9). Cavalerius is well-formed, with a diameter of 40 miles and a central ridge on its floor. The walls rise to 10,000 feet above the interior. It is a fine object when on the terminator.

DE L'ISLE. A 16-mile crater, forming a pair with the less regular 13-mile DIOPHANTUS, to the south. Both craters have central peaks. Various domes lie in the area, and are worth studying.

ENCKE. A crater which may be regarded as the 'twin' of Kepler; but it is a dissimilar twin, since it is far less bright and is not a ray-centre. The diameter is 20 miles, and the walls are rather low. There is no central peak, but a ridge lies on the floor. To the west may be seen the unremarkable formation MÖSTLIN.

GRUITHUISEN. A bright crater on the Oceanus Procellarum, 10 miles in diameter. The area between it and Aristarchus is of great interest. There are the small, bright craters ÅNGSTRÖM and WOLLASTON and the rather larger, less regular KRIEGER, together with rills and domes. The incomplete ring PRINZ has domes on its floor, and there are others nearby. The HARBINGER MOUNTAINS do not form a proper range, but are made up of groups of hills, the highest of which rises to 8,000 feet; all the same, it is possible that the Harbingers once formed part of the border of the Mare Imbrium, perhaps connecting the modern Carpathians with the Jura Mountains.

HERCYNIAN MOUNTAINS. A limb-range near Otto Struve. Some of the peaks here may exceed 7,000 feet in altitude.

HERODOTUS. The companion-crater to Aristarchus. It is 23 miles across, and has a darkish floor, in striking contrast to the brilliance of its neighbour. Its walls rise to about 4,000 feet, but the shape of the crater is not quite regular. Issuing from it is the grand valley, usually called SCHRÖTER'S VALLEY in honour of its discoverer—though this may lead to some confusion, as the crater named after Schröter is a long way away (Section 12). The Valley has been described in the text.

HEVEL. A magnificent 70-mile crater in the Grimaldi chain. Its walls are almost linear in places, but rise to 6,000 feet above a decidedly convex floor which contains a central mountain. Hevel is noted for the system of rills inside it, superbly shown on space-probe pictures but very prominent even from Earth.

HORTENSIUS. A deep 10-mile crater in the Oceanus Procellarum, well-formed and with rather bright walls. To the north may be seen a group of domes; some of the domes have summit pits.

KEPLER. In diameter (22 miles) Kepler is very like Encke, but it is far brighter, and is the centre of one of the most prominent ray-systems on the Moon. There is a central mountain, and the walls are so heavily terraced that they seem to be double in places. To the south-east is a small, deep craterlet, Kepler A. Like Aristarchus, Kepler has interior radial bands, though they are not nearly so pronounced as those in Aristarchus or a few other craters. Kepler has been the site of several reported lunar events.

KUNOWSKY. A small crater, 12 miles across, south-east of Encke. Its floor contains a rather low central ridge.

MARIUS. A well-marked crater 22 miles in diameter, on the Oceanus Procellarum. It has a low central hill, as well as bright streaks on its floor. Well to the south-west is its 'twin', REINER, slightly smaller (diameter 20 miles), with brightish walls but a dark floor.

MAYER, TOBIAS. A crater in the Carpathian Mountains, 22 miles across, and with a central hill; adjoining it to the east is Tobias Mayer A, which is smaller but which also has a central hill. There are various rills and domes in this region.

MILICHIUS. A small bright crater 8 miles in diameter, on the Oceanus Procellarum, west of Copernicus. West of it lies a magnificent dome with a summit pit, and there are various other domes not far off.

OLBERS. A 40-mile crater north-west of Cavalerius; between the two is a smaller crater, B. Olbers is a major ray-centre, and like all ray-craters is very bright. Various other formations beyond it come into view under conditions of extreme libration.

PRINZ. The famous partial ring in the Harbinger Mountains. It has been described with Aristarchus.

PROCELLARUM, OCEANUS. The vast 'Ocean of Storms' has an area of two million square miles, much larger than our Mediterranean, but it is not one of the well-formed circular seas; its surface is lighter and patchier than that of the Mare Imbrium. It extends on to Sections 5 and 9, and joins the Mare Nubium, Mare Humorum, Mare Imbrium and Sinus Roris. Under adverse libration the Oceanus spreads almost to the limb of the Moon, though none of it lies on the far hemisphere.

REINER. Reiner is not unlike Marius, and has been described with it. Various small craters lie on the Oceanus to the south-west; SUESS is the largest of them.

SELEUCUS. This may be regarded as the twin of Briggs, to which it is connected by ridges. It is 32 miles in diameter, with terraced walls rising to 10,000 feet above an interior which contains a central peak. North-east of Seleucus is the distinct 18-mile crater SCHIAPARELLI; a light streak runs south-westward from it, so that Schiaparelli is easy to find under a high light.

STRUVE, OTTO. A vast enclosure made up of two old rings, each about 100 miles across, which have now merged; the west wall is really a ridge parallel with the mountains beyond. In 1952 I discovered a vast crater, only visible under ideal conditions of libration and lighting, beyond Otto Struve and the extreme limb; it has high walls, a central crater and a long ridge on its floor. It had not been previously recorded because it is so seldom visible from Earth in its true guise. It was originally named 'Caramuel', but the IAU Commission altered this name to

EINSTEIN. Photographs of Einstein from space-probes show it to be a truly magnificent formation.

SVEN HEDIN. A large 60-mile irregular formation, with broken walls, between Hevel and Cavalerius on the one side and the limb on the other. There is considerable floor-detail. Sven Hedin is an interesting structure; Olbers, to its north, is the best guide to it.

VASCO DA GAMA. Though 50 miles in diameter, and with a central ridge rising to a peak at its mid-point, Vasco da Gama is too foreshortened to be well seen from Earth. It lies north of Olbers and west of Cardanus and Krafft; Einstein is situated to its north-west.

THIRD (SOUTH-WEST) QUADRANT

The Moon's Third Quadrant includes most of the tremendous Mare Nubium as well as the Mare Humorum and a small part of the Oceanus Procellarum, but much of the quadrant is composed of upland. There are mountains along the limb, and of these the so-called Cordillera and Rook ranges are really parts of the complicated system of the Mare Orientale—something which could never be known before the age of space-probes. There are also the D'Alembert Mountains, but there is some risk of confusion here, as the name has been left off the official maps and the name transferred to a crater on the Moon's far side. Also omitted is the name of the 'Percy Mountains' formerly used for the lofty uplands bordering the Mare Humorum to the west.

Major craters and walled plains in this quadrant include Grimaldi, Riccioli, Schickard, the Walter and Ptolemæus chains, Bailly, Clavius, and of course Tycho. There are notable rill-systems associated with Hippalus, Hesiodus and Sirsalis, and among other features we have the celebrated plateau Wargentin and the remarkable fault mis-called the Straight Wall. Various lunar vehicles have come down in this quadrant, and the region of Mare Nubium in which Apollo 12 landed in 1969 is often called the Mare Cognitum or Known Sea.

Section 9

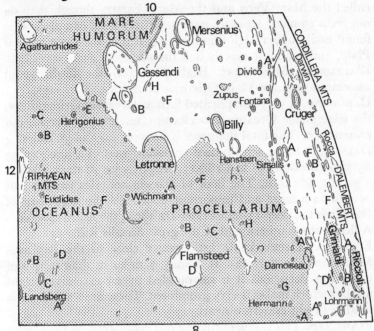

Section 9

AGATHARCHIDES. An irregular formation about 30 miles across, with walls which are of fair height in places (up to 5,000 feet) but which have been almost levelled in others. There is the remnant of a central mountain. The border of the Mare Humorum between Agatharchides and Gassendi has been destroyed—assuming, of course, that it ever existed!

BILLY. This and HANSTEEN make up a pair on the edge of the southern part of the Oceanus Procellarum. Each is about 32 miles in diameter, with walls rising to between 3,000 and 4,000 feet, but Billy is notable because of the darkness of its floor, making it readily identifiable under high illumination.

CORDILLERA MOUNTAINS. As we now know, these and the ROOK MOUNTAINS make up part of the ring boundaries of the Mare Orientale, so that they extend well on to the far hemisphere.

CRÜGER. A 30-mile crater with a very dark floor, resembling

that of Billy. Near it are two small dark plains which have been called the Mare Veris and the Mare Æstatis, though they do not seem to merit separate names. Crüger itself can always be found easily under high illumination; it is of the same type as Plato, and is almost exactly half the size.

D'ALEMBERT MOUNTAINS. High peaks on the limb, again associated with the Mare Orientale complex.

DAMOISEAU. A very complicated formation east of Grimaldi, on the edge of the Oceanus Procellarum, made up of several old crater-rings. The total diameter is between 20 and 30 miles.

DARWIN. A large, semi-ruined enclosure west of a line joining Byrgius (Section 10) to Crüger; Crüger, with its dark floor, is a good guide to it. The floor of Darwin contains various rills as well as a large and important dome. To the north-east of Darwin is the deep 15-mile crater DE VICO; the nearby De Vico A lies at the southern end of the great Sirsalis Rill.

EUCLIDES. A remarkable little crater close to the Riphæan Mountains (Section 12). It is only 7 miles in diameter, and 2,000 feet deep, but it is surrounded by an extensive bright nimbus which makes it very prominent. There are several much smaller bright craterlets nearby.

FLAMSTEED. This is one of the regions considered by G. Fielder to be a 'developing' ring; there is a bright 9-mile crater on the Oceanus Procellarum, north of Letronne, and a 60-mile ghost ring which I personally regard as being ancient. The ghost has incomplete and very low walls, but it is quite distinct under high light.

FONTANA. A 30-mile crater east of Crüger, with low but bright walls and a central hill. There are various rills in this region, which is of course close to the extensive Sirsalis rill-system.

GASSENDI. This is one of the most important formations on the Moon. It is 55 miles in diameter, and lies on the north border of the Mare Humorum. The wall is reasonably high to the east and west, but to the south it contains numerous passes, and has obviously been badly damaged by the Mare material, while to the north the wall has been broken by a prominent, well-formed crater, Gassendi A (named 'Clarkson' on some former maps). The floor of Gassendi includes a central peak, and a magnificent system of rills, shown to advantage on probe pictures but also easily seen from Earth. Gassendi is one of the most 'event-prone' areas on the Moon, and various transient phenomena have been seen there in recent years—which is hardly surprising, since it lies on the edge of a regular sea (Humorum) and is also so rich in rills.

GRIMALDI. The famous dark-floored walled plain near the west limb. It is 120 miles in diameter, so that it is one of the largest enclosures on the Moon, and its floor is so dark-hued that Grimaldi is always unmistakable. The walls are discontinuous, but include some peaks which exceed 8,000 feet. The ramparts are extremely complex, and include hills, ridges, and rills near their foot. The chief feature on the floor is the well-marked crater B. Many TLP have been reported in and near Grimaldi, and gaseous emissions have been detected spectroscopically.

HANSTEEN. This is the companion-crater to Billy, and has been described with it. Unlike Billy, it has a fairly bright floor.

HERIGONIUS. A 10-mile crater on the Mare, north-east of Gassendi. It has rather bright walls, and a central hill.

HUMORUM, MARE. A small part of this interesting Mare is shown in the present Section, but most of it lies in Section 10.

LETRONNE. A good example of a bay. It has a diameter of 70 miles, and borders the Oceanus Procellarum, north-west of Gassendi. Its north wall has been destroyed, and the floor is fairly smooth, though it contains the wreck of a central peak. North of Letronne is a small, distinct craterlet, A.

LOHRMANN. A 28-mile crater lying between Hevel (Section 8) and Grimaldi, so that it is a true member of the celebrated chain. It has a darkish floor, on which is a central hill, and there are many rills nearby. Running obliquely in this region is a curious valley, not well shown on many of the probe pictures, but quite striking under suitable illumination; it was studied by the Japanese astronomer Miyamori, and is known unofficially as the Miyamori Valley. East of Lohrmann, on the Oceanus, is the bright 10-mile crater HERMANN.

MERSENIUS. An important and interesting crater, 45 miles across, near the border of the Mare Humorum. Its walls are terraced, rising to about 8,000 feet in places, and the floor is markedly convex; Hevel is another large formation with this peculiarity. Mersenius is associated with an extensive rill-system; some of the rills lie on the Mare Humorum, others to the south of Mersenius (Section 10) in the direction of Liebig and De Gasparis.

PROCELLARUM, OCEANUS. A small part of the Oceanus extends into this Section, and includes the craters Wichmann, Flamsteed and Hermann as well as the great bay Letronne.

RICCIOLI. The smaller companion of Grimaldi. It is 100 miles in diameter, and has one patch on its floor which is almost as dark as any area in Grimaldi. The interior contains much fine detail.

ROCCA. A large crater, 60 miles in diameter, with irregular walls. It lies south of Grimaldi and north of Crüger.

SIRSALIS. A 20-mile crater which overlaps its slightly larger neighbour Sirsalis A, so that the two form a striking pair similar to that of Steinheil in the Fourth Quadrant (Section 14). Sirsalis, the intruding formation, is much deeper than its twin. Nearby is the famous rill, visible with any small telescope when well placed, which extends from the border of the Oceanus Procellarum southward as far as Byrgius (Section 10).

WICHMANN. A bright 8-mile crater on the Oceanus Procellarum. It is associated with a ghost crater which is even more ruined than that of Flamsteed.

ZUPUS. A very low-walled formation only about 12 miles across, south of Billy. It is easy to find, because its floor, like those of Billy and Crüger, is extremely dark.

Section 10

BYRGIUS. A low-walled and rather obscure enclosure about 40 miles in diameter, not far from the limb. It is easy to find because the small crater on its eastern crest, Byrgius A, is a ray-centre, and so is prominent under high light. Byrgius lies north-west of Vieta.

CAVENDISH. A formation 32 miles in diameter. Its walls are of fair height, attaining 7,000 feet in places, but are disturbed by smaller craters. It lies south-west of Mersenius (Section 9) and west of the Mare Humorum. Between it and the Mare border are various small craters, of which the most prominent are LIEBIG and DE GASPARIS.

DOPPELMAYER. Another splendid lunar bay, this time on the edge of the Mare Humorum. The 'landward' wall is quite high, and there is a central mountain, but the 'seaward' wall has been

Section 10

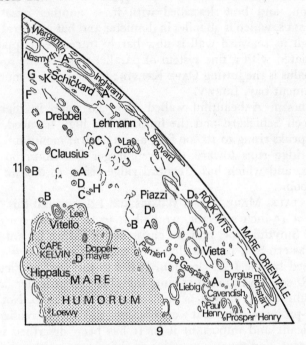

9

so ruined that it is now very low and discontinuous, with wide gaps. Adjoining it is LEE, another incomplete formation which has been ruined by the Mare lava, but which is much less impressive than Doppelmayer. PALMIERI, roughly between Doppelmayer and Vieta, is a curious enclosure whose floor is crossed by several rills.

EICHSTÄDT. A 32-mile, regular-walled crater north-west of Byrgius, and not at all prominent. The Mare Orientale lies beyond.

FOURIER. A 36-mile crater close to Vieta. Its walls are terraced, and the floor contains a central crater in lieu of a central peak.

HUMORUM, MARE. A superb example of a minor circular sea; its area, amounting to 50,000 square miles, is about the same as that of England. It provides probably the most striking example of faulting on the Moon, and there are also many rills and bays in and near it. The floor is relatively featureless, and the only crater on it dignified by a separate name is the very

271

low-walled ghost PUISEUX, near Doppelmayer. Mainly in this Section, and best described with it, is another great bay, HIPPALUS, which is 38 miles in diameter and has a central peak, though its seaward wall is now barely traceable. Hippalus is associated with a fine system of parallel rills. To the south of Hippalus is the jutting CAPE KELVIN; to the north a much less prominent bay, LOEWY.

INGHIRAMI. A beautiful walled plain 60 miles in diameter, between Schickard and the limb. It has high terraced walls, with peaks rising to 10,000 feet, and a central mountain. From it a ridge runs toward the even larger BOUVARD, 80 miles across, and which has a central ridge rising to a peak at its mid-point.

ORIENTALE, MARE. When Wilkins and I discovered this Mare, using a 15-inch reflector, we took it to be a small sea of the Mare Smythii type, and we christened it 'Mare Orientalis'—the Eastern Sea, since at that time east and west had not been reversed by the IAU decision. It lies to the west of Eichstädt and Rocca, and since it is quite invisible except under favourable libration it cannot be shown on the main map. Not until space-probe pictures of it were obtained did anyone realize how significant and important it is. It has been described in the main text, and there is no point in giving topographical details here, because it is so hard to see and because it extends right on to the far hemisphere. The ROOK MOUNTAINS and CORDILLERA MOUNTAINS form its outer and inner ring-wall respectively, which was something else which we could not possibly know when we found it; there is also a fine, regular, central-peaked crater, Schlüter, virtually inaccessible to Earth-based observers. Mare Orientale has proved to be immensely complex. Further on the disk are the large formations LAGRANGE and PIAZZI.

SCHICKARD. This is one of the Moon's greatest walled plains, with a diameter of 134 miles. The walls are rather low, averaging less than 5,000 feet and with its highest peaks rising to just over 8,000 feet. The floor contains some dark areas, as well as various hills and craterlets. Adjoining it to the north-west is the 28-mile LEHMANN, whose floor is connected by passes with that of Schickard. Not far off is the 18-mile, well-formed crater DREBBEL, as well as the small but quite prominent CLAUSIUS and the ill-defined, rather dark-floored LACROIX. The most interesting of Schickard's neighbours is, of course, Wargentin, described separately, while Phocylides and Nasmyth (Section 11) are also members of the Schickard group.

VIETA. A 50-mile crater. Its walls are irregular in height, but in places rise to 15,000 feet; there is a minor central peak.

VITELLO. This is a splendid example of a concentric crater—even though the complete inner ring is not quite concentric with the main wall. Vitello is 30 miles across, and has a central peak crowned by a craterlet. It lies on the border of the Mare Humorum, east of Doppelmayer and Lee, and its seaward wall has been clearly reduced by the Mare lava, though elsewhere the rampart rises to over 4,000 feet above the interior.

WARGENTIN. This is one of the most remarkable formations on the Moon, and represents the only example of a really large, well-preserved lunar plateau. It is 55 miles in diameter, so that in size it is the equal of Copernicus, and adjoins Schickard. There are various hills and ridges on the plateau surface. There is a 'wall' in places, but the whole floor is raised by about 1,400 feet. Despite its unfavourable position close to the limb, it is easy to find, and is well worth careful study.

Section 11

BAILLY. The largest formation on the Earth-turned hemisphere which is officially classed as a walled plain. Its area is more than half that of the Mare Humorum, but it has a light floor. It has been described as 'a field of ruins'; even though peaks in its walls rise to about 14,000 feet, the height of the rampart is very irregular. The floor is crowded with detail, notably one large, well-formed crater, B. West of it, even nearer the limb, is HAUSEN; and along the limb toward the south pole there are various quite major formations such as LEGENTIL, DRYGALSKI, CABÆUS and MALAPERT, which are difficult to see because of extreme foreshortening. Before the space-probe pictures became available, this whole region was very imperfectly mapped. Beyond Bailly lie some peaks which have been called the DÖRFEL MOUNTAINS, though it is now known that they do not make up a lofty, continuous range, as was once thought likely.

273

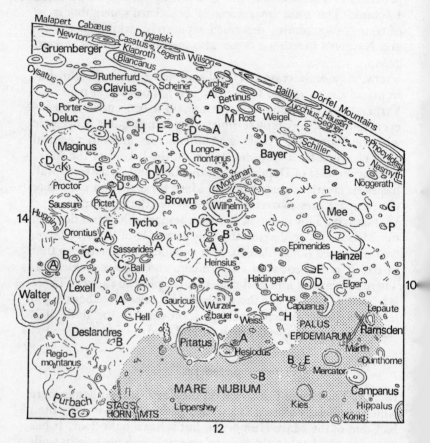

12

BLANCANUS. This and its companion SCHEINER are rather
dominated by their vast neighbour Clavius, but both are major
formations in their own right. Blancanus is 57 miles in diameter,
Scheiner 70; both have lofty walls, rising to 12,000 and 15,000
feet respectively; both have much interior detail—a nearly
central craterlet in the case of Scheiner. Close to the limb near
here are several prominent craters, WILSON, KIRCHER, BETTI-
NUS, ZUCCHIUS and SEGNER, which are more or less lined up and
were probably formed along the same line of weakness in the
lunar crust. All five are between 40 and 50 miles across, with

lofty, continuous walls. The area between this crater-line and Scheiner seems to be occupied by what looks like an old ring, though its walls have been completely destroyed.

CAMPANUS. A well-formed, 30-mile crater at the edge of the Mare Nubium and the Palus Epidemiarum. It is the twin of Mercator, but its floor is lighter, though still on the dusky side. There is a central hill. Various rills run between Campanus and Hippalus, associated with the Hippalus system. To the south-west is DUNTHORNE, which has rather broad walls.

CAPUANUS. This is an extraordinary formation. It is 35 miles across, and lies on the edge of the Palus Epidemiarum; its floor has been to some extent flooded, and appears darkish, while the walls have been disturbed on the seaward side and badly ruined in places. What makes Capuanus so notable is the fact that at least eight major domes lie on its floor; some of them are visible with small telescopes. There is no other known case of a large crater so rich in domes. Closely outside Capuanus is an imperfect formation, ELGER.

CASATUS. This and KLAPROTH form another example of overlapping craters—this time on a grand scale, since Casatus, the smaller of the two, is a full 65 miles in diameter, with high walls. Klaproth is shallower, with a much smoother floor. Not far from Klaproth, to the west, are several well-formed craters. Drygalski, already referred to, lies in the libration region beyond Casatus.

CICHUS. A prominent 20-mile crater east of Capuanus, just beyond the border of the Mare Nubium. On its western crest is a well-formed crater, Cichus G, about 5 miles in diameter. To the south is the distinct Cichus A, and to the north of Cichus is WEISS, which is so broken that it gives the impression of being an enclosure bounded by irregular ridges rather than a true crater. There are various rills in this area, no doubt associated with the Hesiodus system.

CLAVIUS. Apart from Bailly, Clavius is the largest of the so-called walled plains, since it is 145 miles in diameter, with mighty walls rising to over 12,000 feet. Every lunar observer knows it well. The north-east wall is broken by a large crater, PORTER, and there is a chain of craters arranged in an arc across the floor, of which RUTHERFURD is the largest. When on the terminator, Clavius is distinctly visible with the naked eye, and is easy to find telescopically under any conditions of illumination.

DELUC. An unremarkable 28-mile crater south-east of Maginus.

DESLANDRES. A huge enclosure west of Walter and Regiomontanus; on older maps it was called Hörbiger (a name due to the German selenographer Philipp Fauth, a devotee of the extraordinary ice theory proposed by Hans Hörbiger, and described in the text). Deslandres is distorted in outline. The regular crater HELL lies inside it; on the border are BALL, which is 25 miles across and has high terraced walls, and the 39-mile LEXELL, whose north wall has been reduced to such an extent that Lexell now resembles a bay opening out of Deslandres, even though it retains the wreck of a central peak. In the north part of the floor of Deslandres is a low-walled crater, B.

EPIDEMIARUM, PALUS. A conspicuous dark plain extending out of the Mare Nubium. Mercator and Campanus lie on its borders. It is notable mainly because there are many rills in and near it, known as the Ramsden system even though the crater of RAMSDEN itself is only 15 miles in diameter. In complexity, the Ramsden system rivals that of Triesnecker. Closely west of Ramsden is another small, unremarkable crater, LEPAUTE.

GAURICUS. An irregular enclosure with a diameter of about 40 miles, and walls which are uneven in height. It is a member of the Pitatus group; Wurzelbauer is the third member.

GRUEMBERGER. This and its smaller companion CYSATUS lie south-west of Clavius, and belong to the Moretus group (Section 14).

HAINZEL. A curious formation, made up of two rings which have coalesced; the north-south diameter is 60 miles. Under oblique lighting it is conspicuous, but it is hard to find near full moon. It lies north of Schiller. East of it is a small crater, EPIMENIDES, and to the south is a large ruined enclosure, MEE, with low walls.

HEINSIUS. A very peculiar structure between Tycho and Capuanus, rather closer to Tycho. The north wall is quite high, but in the south the rampart has been broken by three considerable craters, one of which really lies on Heinsius' floor. The diameter of Heinsius itself is about 45 miles.

HELL. A 20-mile crater with a low central hill, lying near the western edge of the great enclosure Deslandres.

HESIODUS. This is the companion of Pitatus, to which it is connected by passes in their common wall. It is 28 miles across, and its walls have been somewhat reduced by the Mare lava; the floor contains an almost central crater. From Hesiodus a famous rill runs across to the mountain arm north of Cichus; it is easy to see with a small telescope when well placed. Other, less prominent rills are associated with it.

KIES. This cannot be termed a ghost-ring, but it lies on the Mare Nubium, and its walls are now very low, nowhere exceeding 2,500 feet; the floor is flooded by lava. West of it lies a superb example of a dome with a summit pit, and there are other domes in the area. Other smaller craters nearby are KÖNIG, Kies A, and Agatharchides A, in which is a dark interior radial band, and which lies on one of the Hippalus rills.

KLAPROTH. This large crater has been described with its companion, Casatus.

LIPPERSHEY. A 4-mile crater west of the Stag's-Horn Mountains, which actually intrude into this Section from Section 12.

LONGOMONTANUS. A very large enclosure, 90 miles in diameter, with complex walls and considerable detail on its floor. It is easy to find, though not nearly so prominent as Clavius. North of it, between Longomontanus and Wilhelm I, is MONTANARI, with two distinct craters on its west wall, which is common to another rather ill-defined walled plain, LAGALLA.

MAGINUS. With a diameter of 110 miles, Maginus is one of the grandest walled plains on the Moon, and would seem even more striking were it not so close to the even more majestic Clavius. Maginus has walls of irregular height, and a rather rough floor. To the south-west the rampart is broken by a 30-mile crater, Maginus C; another crater of about the same size, PROCTOR, lies outside the north wall. Oddly enough Maginus is difficult to identify near full moon—even if it does not vanish entirely, as some books maintain.

MERCATOR. The twin of Campanus, on the edge of the Palus Epidemiarum. It has a dark floor, with walls rising in places to 5,000 feet; there is only a trace of a central peak, but the floor contains some detail, including a delicate rill. To the south-west is MARTH, which has a complete inner ring and is an excellent example of a 'concentric crater'.

NEWTON. This is generally regarded as the deepest of the lunar walled plains. It appears to be compound, but is so foreshortened that to Earth-based observers it is a difficult object to study. It lies south-west of Moretus (Section 14) and east of the Casatus-Klaproth pair. Between Newton and Moretus is another crater, SHORT, which also lies in Section 14.

NUBIUM, MARE. Part of the Mare Nubium is shown here, but most of it lies in Section 12.

ORONTIUS. An irregular formation about 52 miles in diameter, north-east of Tycho. It is one of a group which includes Miller and Nasireddin in Section 14, and HUGGINS and SAUSSURE in this Section. Orontius has been disturbed by Huggins; the whole area is crowded with detail, but includes no particularly notable features.

PHOCYLIDES. A most interesting formation, 60 miles in diameter. It is a member of the Schickard group (Section 10), which also includes NASMYTH (between Phocylides and Wargentin) and Wargentin itself, as well as Phocylides C, to the north-west. There is considerable detail on the floor of Phocylides, and there are indications of a major step-fault there. Between Phocylides and the limb lies a regular crater, PINGRÉ; and between Phocylides and Mee there is another crater, NÖG-GERATH, which is unremarkable but distinct.

PITATUS. A grand formation 50 miles across. It lies on the border of the Mare Nubium, and gives the impression of a large lagoon; its walls have been badly damaged, and are very low in places. Passes connect the floor with that of Hesiodus. Pitatus has no true central peak, but there is a hill not quite in the middle of the floor. The other members of the Pitatus group are Gauricus and Wurzelbauer.

PURBACH. This great ring-plain, 75 miles in diameter and with walls rising to 8,000 feet in places, is a member of the chain which includes Regiomontanus and Walter. There is considerable detail on the floor. The outline of Purbach is not entirely regular, and the northern part of the rampart has been disturbed by later outbreaks. To the north-west lies a fairly regular crater, LACAILLE, which comes between this Section and Section 14; it is also adjacent to Blanchinus (Section 14).

REGIOMONTANUS. A distorted formation between Purbach and Walter. Its east-west diameter is 80 miles, but the north-south diameter is only 65 miles; one has the impression that the whole crater has been 'squashed' between Purbach and Walter. There is abundant floor-details, including some peaks near the centre. The walls are of irregular height, with some peaks reaching 7,000 feet. Regiomontanus touches both Purbach and Deslandres.

SASSERIDES. An irregular enclosure north of Tycho, with a diameter of about 60 miles. Its north wall has been largely destroyed by four smaller craters, and the floor includes numerous pits and hills. To the south-west is Sasserides A, which has a central peak, and between Sasserides and Orontius is a regular formation, Orontius A.

SCHEINER. This great plain may be regarded as the twin of Blancanus, though it is the larger and deeper of the two. It is described with Blancanus.

SCHILLER. A compound formation in the area between Schickard and Clavius. It is 112 miles long, but only 60 miles wide at its broadest point, and is the result of the fusion of old rings. The floor contains some ridges and pits. The region between Schiller, Segner and Phocylides is probably a very old ring whose walls have been destroyed. Closely outside Schiller is the well-formed BAYER, 32 miles across, and with high terraced walls attaining 8,000 feet in places.

TYCHO. This crater, 54 miles in diameter, is in a class by itself, as it is the centre of by far the greatest ray-system on the Moon; it has been fully described in the text, and the last vehicle of the Surveyor series now stands on its outer slopes. Tycho's neighbours include PICTET, BROWN and STREET as well as Sasserides and the Orontius group.

WALTER. A majectic, complex walled plain 90 miles in diameter, with an asymmetrically-placed interior mountain and several considerable craters on its floor. It is the senior member of the trio which includes Regiomontanus and Purbach.

WILHELM I. A 60-mile walled plain with walls which are irregular in height, but which contain a few peaks reaching 11,000 feet above the floor. Outside it, to the south-west, is the irregular, rather pear-shaped Lagalla; to the north-west is a deepish crater, D, and beyond this, in a northerly direction, is

Heinsius. Wilhelm I is easy enough to find, as it lies not far from Tycho, but it does become very obscure under high illumination.

WURZELBAUER. This is one of the Pitatus group, and is about 50 miles in diameter, but its walls are irregular, and very low in places. The floor contains a mass of complex detail.

ZUCCHIUS. This is one of the chain which includes Segner, Bettinus, Wilson and Kircher. It has been described with Blancanus. Between Zucchius and Longomontanus lie two unremarkable but quite distinct craters, WEIGEL and ROST.

Section 12

ALPETRAGIUS. Though only 27 miles in diameter, Alpetragius is a splendid sight when observed under good conditions. It lies closely outside the walls of Alphonsus and Arzachel, and is distinguished for its very high terraced walls, which exceed 12,000 feet in places, and for its enormous central mountain, which is rounded and which is crowned by two summit pits.

ALPHONSUS. Little more need be said about this great formation here, since it has been so fully described in the main text. It achieved particular notoriety in 1958 when Kozyrev saw a red event inside it, but for many years before that it had been regarded as a probable site of mild activity. It is the middle member of the Ptolemæus chain, and is slightly distorted in form; on the floor there is a minor central mountain and a mass of detail, including a fine system of rills. It also contains the remains of Ranger 9, though no telescope on earth will show it! Obviously, Alphonsus should be kept under close surveillance, since further TLP may occur in it at any time.

ARZACHEL. The southern member of the Ptolemæus chain. It is 60 miles in diameter, and has high walls, reaching 13,500 feet in places; the floor includes an elongated central mountain and a deep, prominent crater, Arzachel A, well over to the west of the peak. The gradation in type from Ptolemæus to Alphonsus and then Arzachel is very interesting and significant. There can be no doubt that all three were formed by the same process, which adds force to my contention that there is no basic difference between a 'walled plain' such as Ptolemæus and a true crateriform structure such as Arzachel—or, for that matter, Alpetragius.

Section 12

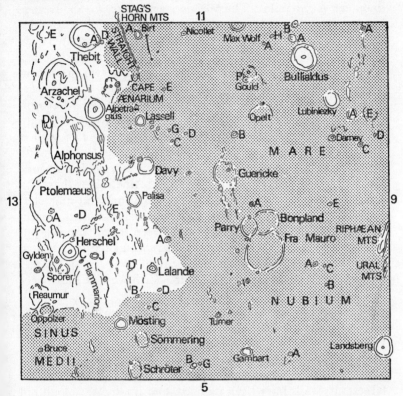

BIRT. A very interesting little crater 11 miles in diameter, on the Mare Nubium to the west of the Straight Wall. Its south-east wall is disturbed by a smaller crater, A, and at the junction of the walls there is a still smaller formation. The walls and profile of Birt are irregular, and there are two dusky bands running across the floor to the western wall; these show optical changes over a lunation. Closely outside, to the west, is the famous rill, which is in part crateriform. Birt may be found even under high light, when the Straight Wall is completely invisible.

BONPLAND. This is a member of the Fra Mauro group on the Mare Imbrium, and is described with Fra Mauro.

BULLIALDUS. A particularly fine crater; it has been described as a miniature Copernicus, though it is not a ray-centre. The diameter is 39 miles. The massive walls rise to an average height of over 8,000 feet above a floor which includes a complex central mountain group; the inner slopes of the rampart are superbly terraced. Bullialdus is, in fact, one of the most perfect of all formations of its type. To the south are two prominent deep craters, Bullialdus A and B, with König (Section 11) some way to the west of B, between Bullialdus and Mercator.

DAVY. A 20-mile crater near the edge of the Mare Nubium, between Alphonsus and the Fra Mauro group. The wall is quite high in places, but in the south-east it is broken by a small, deep crater, Davy A. To the south is the low-walled 14-mile crater LASSELL, and between Lassell and Alpetragius there is a very symmetrical bright crater, Lassell B, 6 miles in diameter. Some way north-west of Davy there is a low-walled, rather incomplete ring, PALISA; the area between it and Davy seems to represent the ruins of a very battered and reduced walled plain considerably larger than Davy itself.

FLAMMARION. An irregular enclosure north-west of Ptolemæus and Herschel, with a maximum diameter of 45 miles. Its walls are incomplete, particularly in the north, where they have been reduced by material from the adjacent mare surface.

FRA MAURO. This is the largest member of a group of craters in the Mare Nubium; the others are Bonpland, Parry and Guericke. Fra Mauro itself, with a diameter of 50 miles, is the largest; its walls have been reduced, and it almost, though not quite, comes into the category of a ghost. Bonpland (36 miles across) is in a slightly better state of preservation, and Parry (26 miles) has still higher walls. These three formations have common boundaries, and are crossed by various rills. Further south is GUERICKE (sometimes, though wrongly, spelled Guériké) with incomplete walls, very broken in the north and almost levelled in places. The interior contains considerable detail, including a crater-chain, pits and ridges, several delicate rills, and one distinct craterlet, D. Apollo 14, with Astronauts Shepard and Mitchell, came down in the Fra Mauro area, so that information is still being received from there.

HERSCHEL. A fine crater 28 miles across, with terraced walls and a large central peak. It lies closely north of Ptolemæus. Adjoining it to the north is SPÖRER, much less complete, and showing some evidence of having been partly filled with lava. To the west of Herschel lies GYLDÉN, low-walled but reasonably regular, and in the triangle formed by Herschel, Spörer and Gyldén there is a splendid example of a crater-valley, easy to see with any small telescope when suitably lit.

LALANDE. A well-formed crater 15 miles in diameter, with a low central hill. Between it and Flammarion there is a very old incomplete ring, on the south wall of which is a distinct crater, D, not much smaller than Lalande itself, and of the same type.

LANDSBERG. The more correct spelling is 'Lansberg', but the alternative form has been in use for so many years that to alter it now would be pedantic. The crater lies on the Mare Nubium, north of the Riphæan Mountains, and is a fine example of a ringed plain, 28 miles in diameter, with massive walls and a central mountain. Well to the west lies the shallower and less regular GAMBART; between Gambart and Landsberg is a distinct little crater, Gambart A; and between Gambart and Lalande there are two craterlets, TURNER and Gambart F.

LUBINIEZKY. A very reduced crater on the Mare Nubium, north-west of Bullialdus; its walls are everywhere low, and in places discontinuous, while the floor is as dark as the surrounding Mare. To the north lies the distinct little crater DARNEY. There are numerous ghost rings in this area; GOULD and OPELT are others.

MEDII, SINUS. The Central Bay. Part of it is shown in this Section; Mösting, Sömmering and Schröter lie near its boundary. On the Sinus itself are two distinct craterlets, BRUCE and BLAGG.

MÖSTING. A well-formed crater 16 miles across, with a low central hill. To the south-south-east is Mösting A, which is very bright and is a minor ray-centre; it has been used as a reference point for various lists of the positions of lunar features. It actually lies on the wall of Flammarion.

NICOLLET. A distinct 10-mile crater on the Mare Nubium, roughly west of Birt. Some distance away is MAX WOLF, low-walled and irregular in outline.

NUBIUM, MARE. One of the largest of the lunar seas; much of it lies in this Section, though parts extend into Sections 9 and 11. It is lighter, patchier and much less regular than the Mare Imbrium, and there are no really high ranges to mark its border. Craters on it include Bullialdus and the Fra Mauro group. Various space-probes have landed in the Mare Nubium; the region where Apollo 12 came down is often called the 'Mare Cognitum' or Known Sea, though there really seems no justificiation for a separate name.

PARRY. This crater is in the Fra Mauro group, and has been described with it.

PTOLEMÆUS. One of the most famous walled plains on the Moon. It is over 90 miles across, and is near the centre of the disk, so that it is ideally placed for observation. Its floor is darkish in hue, but contains various objects as well as a well-marked craterlet, Ptolemæus A (called 'Lyot' on some former maps, though the name Lyot has now been transferred to a crater on the Moon's far side). The walls of Ptolemæus are fairly continuous, and contain some high peaks, rising to 9,000 feet or so. Ptolemæus is a member of a great chain of walled formations; the other members are Alphonsus and Arzachel.

RÉAUMUR. A low-walled ring near the edge of the Sinus Medii. Its neighbour OPPOLZER is similar. Rhæticus (Section 12) lies south-west.

RIPHÆAN MOUNTAINS. A low mountain range on the Mare Nubium, lying along the wall of a large ghost-crater. Its highest peaks rise to no more than 3,000 feet. The URAL MOUNTAINS, to the north, are really part of the Riphæans. The best means of identification is provided by Euclides (Section 9), whose bright nimbus makes it easy to find under any conditions of illumination.

SCHRÖTER. A semi-ruined, 20-mile crater near the common border of the Mare Nubium and the Sinus Medii. It has been badly reduced by Mare lava, and its walls are incomplete in the south. Note that it is nowhere near Schröter's Valley, which lies in the Second Quadrant near Herodotus.

SÖMMERING. Another reduced ring, 17 miles across, close to Schröter and Mösting. Its walls are low everywhere, with a gap in the south.

STRAIGHT WALL. As has been made clear in the text, this formation is not straight, and is not a wall. It is a particularly interesting fault, between Thebit and Birt, ending to the south in a group of hills often called the STAG'S-HORN MOUNTAINS; these peaks make up part of the boundary of an ancient ring lying to the west. Before Full Moon, the Wall appears as a dark line; for some time around Full it is invisible, and after Full it reappears as a bright line, as the Sun shines upon its inclined face.

THEBIT. A very significant crater; it is broken by a smaller formation, A, which is in turn broken by the still smaller D. It is always identifiable, even when the Moon is full. It may be a true member of the Walter chain, since its outer slopes merge with those of Purbach to the south-west. Some way north-west of Thebit lies the cape known as PROMONTORIUM ÆNARIUM—a mis-spelling of the proper name of Tænarium, but which has become generally accepted.

TURNER. This small crater has been described with Landsberg and Gambart.

FOURTH (SOUTH-EAST) QUADRANT

This quadrant contains few seas; the Mare areas are limited to part of the Mare Fœcunditatis, all of the Mare Nectaris, a very small portion of the Sinus Medii, the irregular Mare Australe on the south-east limb, and a very slight area of the Mare Tranquillitatis—though an important one, since it was here that the Lunar Module of Apollo 11 brought Neil Armstrong and Edwin Aldrin down for that first landing in July 1969.

Most of the quadrant is occupied by rugged uplands, and there are craters and walled plains of all sorts, ranging from the huge, ruined Janssen to superb formations such as Theophilus; there are smaller rings, and almost countless craterlets. Lofty mountain ranges are absent, but there is the interesting Altai Scarp running north-westward from Piccolomini.

Very few transient phenomena have been seen in this quadrant, and the paucity of events seems to be due to something more fundamental than mere observational selection.

Apollo 11 is not the only probe to have landed in this area. Apollo 16 came down in the region of Descartes, while Russia's Luna 16 brought home rock samples from the Mare Fœcunditatis.

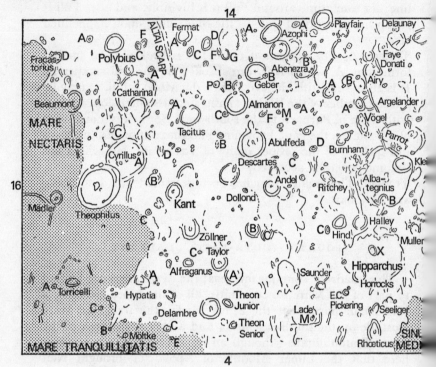

Section 13

ABENEZRA. This crater forms a notable pair with Azophi. The two are of about the same size (27 miles in diameter) and have high walls; the inner ramparts are terraced, particularly with Abenezra. To the south-west, just on Section 14 but shown here also, is PLAYFAIR, equal in size to Abenezra and Azophi, and there is a fairly distinct craterlet, A, between Playfair and Azophi.

The most interesting feature of the whole group is that Abenezra overlaps a shallower formation, C, which is if anything fractionally the smaller of the pair, so that we may have a possible departure from the general rule even though the difference in size between the two is negligible.

ABULFEDA. Another interesting pair is made up of Abulfeda and Almanon, not far from Abenezra; the intervening space contains Geber. Abulfeda is larger and deeper than Almanon, with a diameter of 40 miles and walls rising to 10,000 feet; the figures for Almanon are 30 miles and 6,000 feet. A crater-valley runs from north-west to south-east between them, and extends toward the general direction of Polybius; undoubtedly it is associated with the Altai Scarp. To the north of Abulfeda are DESCARTES and ÅNDËL, which have low walls, as well as the bright 6-mile DOLLOND. Dollond borders a large, ruined enclosure, and some maps give the name to the ruin instead of the deep craterlet. It was in the Descartes area that Apollo 16 landed.

ALBATEGNIUS. A great crater 80 miles in diameter, near the Ptolemæus chain (Section 12) and forming a notable pair with Hipparchus. The walls of Albategnius are generally quite high, with one or two peaks rising to well over 10,000 feet, and there are prominent terraces. The walls are, however, broken in the south-west by a large crater, KLEIN. There is a central mountain on the floor of Albategnius, as well as various craterlets; it is a fine sight under oblique lighting.

ALFRAGANUS. A very conspicuous crater. It is only 12 miles in diameter, but is very bright, and as it is also the centre of a minor ray-system it is easily found at full moon; it lies in the uplands, well north-west of Theophilus and south-east of Delambre. Near it are the triangular, 30-mile HYPATIA, almost on the Mare border; the elliptical TAYLOR, 25 miles across, with a group of three craters to its north-west; and the low-walled, rather irregular ZÖLLNER.

ALMANON. This crater has been described with its companion, Abulfeda.

ALTAI SCARP. This is certainly more of a scarp than a mountain range, and is part of the ring-system of the Mare Nectaris; most of the Mare lies in Section 16. The Scarp rises to an average of 6,000 feet above the general level to the east, but only very slightly above the level to the west. Most of it lies in Section 14, but it extends past Fermat almost as far as Tacitus. FERMAT itself is 25 miles in diameter.

ARGELANDER. A 20-mile crater south-east of Albategnius, with terraced walls and a central peak. Its twin is the nearby AIRY, rather similar to it, and also with a central peak. To the east, a crater-rill runs across into Parrot.

AZOPHI. This adjoins Abenezra, and has been described with it.

BEAUMONT. An excellent example of a bay. It lies on the edge of the Mare Nectaris, between Fracastorius and Cyrillus, and is 30 miles in diameter; it retains its landward wall, but the rampart on the Mare side has been largely destroyed by lava. There are some very small, delicate craters on the floor, but little else.

BURNHAM. A low-walled formation south-east of Albategnius.

CATHARINA. The southernmost member of the Theophilus chain. It is about 55 miles in diameter, with rugged walls; the floor contains a large, low-walled, ruined ring in the north. There is no central peak, so that, as with other trios, we have a gradation in type from Theophilus through Cyrillus to Catharina. The area between Catharina and Cyrillus is high, and is a splendid sight under oblique illumination.

CYRILLUS. Cyrillus lies between Catharina, from which it is separated by the upland area described above, and Theophilus, which overlaps it. As usual, the walls of the broken formation (Cyrillus) remain perfect up to the point of junction, ruling out any violent method of origin. The floor of Cyrillus contains a reduced central hill, as well as a considerable crater, Cyrillus A, and much fine detail. The diameter is about 60 miles. Between Cyrillus and Kant, to the north-west of Cyrillus itself, is a distinct crater, B.

DELAMBRE. A well-formed 32-mile crater west of the Mare area. It has high walls, with some peaks reaching to 15,000 feet. This was the landing area of the unmanned probe Ranger 8, which obtained excellent photographs of Delambre during its plunge Moonward. To the west lie the two Theons, both of which are bright. Delambre contains a central peak with a summit pit.

DELAUNAY. A most peculiar compound formation to the north of Blanchinus, north-east of Purbach (Section 11); Purbach, Lacaille, Dalaunay, Faye and Donati are roughly lined up. Delaunay is made up of two irregular structures with a ridge between them, common to both.

FAYE. This and DONATI are two irregular, imperfect formations near Delaunay. Each is about 22 miles in diameter, and each has a central peak.

FRACASTORIUS. This great 60-mile bay lies partly in this Section and partly in Section 16; it marks the southernmost point of the Mare Nectaris. The seaward wall has been almost destroyed, but is still traceable. There is a darkish streak across the floor which is of a slightly reddish hue, and is detectable with a moonblink device, making a good test for atmospheric conditions of observation. Abutting on it to the west is an irregular, somewhat triangular depression, D.

GEBER. A regular, 25-mile crater between Almanon and Abenezra. Its walls are high and terraced, but disturbed to the west by a much smaller crater, Geber B.

HALLEY. This and HIND are two well-formed, terraced craters close to Hipparchus and Albategnius. Halley is 22 miles in diameter, Hind 16. A valley runs from Halley into Hipparchus in the one direction, and to the eastern glacis of Albategnius in the other. West of Halley, between Hipparchus and Ptolemæus, is the low-walled, irregular formation MÜLLER, and some way south of Hind is another deformed object, RITCHEY.

HIND. This crater has been described with its larger companion, Halley.

HIPPARCHUS. This tremendous enclosure, almost equal to Ptolemæus in size, has been very broken, but is still striking when seen under low illumination. A grid system is very evident in this area, and Hipparchus contains a great amount of fine detail as well as the prominent 18-mile crater HORROCKS. Outside Hipparchus, to the north-east, are the reduced formations SAUNDER and LADE, and the more regular but smaller crater E. C. PICKERING. Hipparchus is, of course, much less imposing than Ptolemæus (Section 12).

KANT. A very deep crater, 20 miles in diameter, west of Theophilus. It is distinguished by its huge, rounded central mountain, which is crowned by a summit pit. Kant is one of the few craters in this quadrant in which a transient event has been reliably reported—by the French astronomer Trouvelot in January 1873; on the 4th of that month he stated that the crater was 'filled with mist'.

MÄDLER. A prominent irregular crater 20 miles across, on the Mare Nectaris east of Theophilus. Its floor is crossed by a ridge, which joins the central mountain mass. Mädler lies near the

K   289

western border of a ghost ring which extends in the direction of Isidorus (Section 16).

MEDII, SINUS. A small part of the Sinus is shown in this Section, in the area of Rhæticus.

NECTARIS, MARE. Part of the Mare Nectaris appears here; the rest is in Section 16. The Mare is basically circular, but has been somewhat distorted, with a good deal of faulting; the Altai Scarp represents part of its system. The only named crater on the Mare is Rosse (Section 16), which is small and rather bright. On the edge of the Mare are Theophilus and the bays of Beaumont and Fracastorius.

PARROT. A very complex, irregular and compound structure, around 40 miles across, south of Albategnius. A crater-valley runs into it from the area of Argelander and Airy.

POLYBIUS. A fairly regular 20-mile crater south-west of Catharina.

RHÆTICUS. This and RÉAUMUR (Section 12), each 28 miles in diameter, lie on the border region of the Sinus Medii. Both are somewhat reduced, with low walls; SEELIGER, between Rhæticus and Hipparchus, is even more distorted.

TACITUS. A somewhat polygonal formation, 25 miles in diameter, between Catharina and Abulfeda. The walls are terraced, and rise in places to 11,000 feet. The floor contains two craterlets, one of which is almost central.

THEON JUNIOR and THEON SENIOR. Two prominent craterlets, respectively 10 and 11 miles in diameter, west of Delambre. Theon Junior is slightly the more brilliant of the two, but both are bright. In many ways they resemble Alfraganus.

THEOPHILUS. There can be little doubt that Theophilus is, with the possible exception of Copernicus, the grandest crater on the whole Moon. It is extremely deep, with walls rising to 18,000 feet above the floor; there is a splendid, many-peaked central mountain mass, and the inner ramparts are terraced. It is always a magnificent sight, and may be recognized under any conditions of illumination. It is the northern member of a chain of three great walled plains, and actually intrudes into its neighbour Cyrillus.

TORRICELLI. A curious, compound formation on the Mare, north of Theophilus. It is made up of two rings; the larger one, to the east, has a diameter of 12 miles. The general impression is that of a pear-shaped enclosure. There are interesting features inside Torricelli, including some delicate rills.

TRANQUILLITATIS, MARE. A small part of this large Mare appears here; the rest lies in Section 4. Near the border is a small, distinct crater, MÖLTKE. This is, of course, the region in which Apollo 11 came down.

VOGEL. A peculiar formation near Albategnius, north of Argelander. It consists of three craters which have merged, so that it may be classed as a short crater-chain. There are various valleys in the neighbourhood. North-east of Vogel lies the obscure crater BURNHAM.

Section 14

ALIACENSIS. A noble crater 52 miles in diameter. It lies outside the wall of Walter, and has a rather smaller 'twin', Werner. Aliacensis has broad, terraced walls and a central mountain. South of it lie two somewhat broken rings, KAISER and NONIUS.

ALTAI SCARP. Much of the Scarp lies in this Section; it begins near Piccolomini, and runs north-westward to Tacitus in Section 13.

APIANUS. A 39-mile crater with high terraced walls, rising in places to 9,000 feet. It lies east of the Aliacensis-Werner pair, and in between it and Werner is a large, broken enclosure which has been called KRUSENSTERN. South-east of Apianus is the very irregular POISSON, which is a compound structure with a mean diameter of about 45 miles.

BIELA. A 46-mile crater rather near the limb, not far from Pontécoulant and the Vlacq group. It has high, terraced walls, and a central peak. To the north-east the wall is disturbed by the intrusion of a considerable crater, Biela A.

BLANCHINUS. A crater north-west of Werner; it is 33 miles in diameter. Its walls are high in places, but are somewhat uneven, and the floor contains much fine detail. Craters near it include Faye and Delaunay (Section 13).

BOGUSLAWSKY. A major formation, 60 miles across and with high walls containing peaks up to 11,000 feet. It is too near the limb to be well seen.

BOUSSINGAULT. Another great formation, along the limb eastward from Boguslawsky. It is interesting inasmuch as it consists of three rings, with a maximum diameter of 70 miles. Even

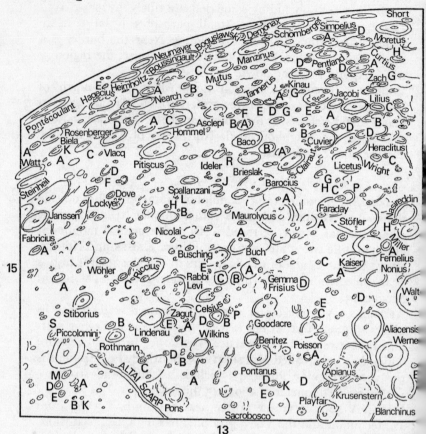

13

worse placed, and seen only under favourable libration, are
HELMHOLTZ (60 miles in diameter) and NEUMAYER (50
miles). In 1954 I detected some curious ray-like features crossing Helmholtz, not like any others known to me; and even after
studying Orbiter and Apollo photographs I am still uncertain
as to their true nature.

BUCH. A regular crater, 30 miles across, north-east of Maurolycus. The floor is relatively smooth, though it includes some
low-rimmed pits. Adjoining it to the north-east is—

BÜSCHING, slightly larger (diameter 36 miles) but less regular.
The region surrounding Buch and Büsching is rich in craterlets.

CURTIUS. A deep 50-mile crater in the south polar region, not far from Moretus, with massive, terraced walls.

CUVIER. A moderately regular, 50-mile crater, with high terraced walls and a central hill. It lies in the uplands, east of the Heraclitus-Licetus group. Close by are CLAIRAUT, which has been deformed by the intrusion of two distinct craterlets, and BACO, 40 miles across, which has lofty walls and a low central peak. North of Baco lies BREISLAK, which is slightly smaller; between the two is a small, deep craterlet. There are two well-formed craters of some size between Baco and Clairaut.

DEMONAX. This great enclosure, 75 miles in diameter, lies between Boguslawsky and the limb, but is visible only under favourable libration. This is a pity, since it is majestic and complex.

FABRICIUS. A walled plain 55 miles in diameter, breaking into the vast ruin Janssen. There is a central mountain on the generally rough floor, and the walls attain 9,500 feet in places. A crater-valley links it with the crater A, to the north, while a long rill runs northward from outside the west wall of Fabricius. To the north-east is Metius (Section 15) which is of about the same size as Fabricius.

FARADAY. A very irregular formation, around 40 miles in diameter, which has broken into Stöfler. There are two deep craters which break into the walls of Faraday itself, and the whole terrain is very complex. The floor of Faraday is much rougher than that of Stöfler.

GEMMA FRISIUS. A large, irregular enclosure north of Maurolycus, much broken by smaller craters, and with a decidedly rough floor. In the north it is disturbed by the rather more regular, 30-mile GOODACRE, which has a low central peak.

HAGECIUS. A most peculiar formation; it is a member of the Vlacq group, south-west of Janssen. Its diameter is about 50 miles, but one wall has been ruined by the intrusion of no less than five craters. The floor is relatively smooth, but contains several craterlets.

HERACLITUS. A strange, irregular enclosure, with a central ridge. It adjoins LICETUS, and Cuvier may also be regarded as a member of the group. Licetus is 46 miles across, with uneven walls and a low central hill; the rampart has been broken on the south, so that the interior connects with that of Heraclitus, though Cuvier remains separate. The whole group can be very prominent when well-placed, and is not difficult to find, since it lies not far south of the dark-floored and always prominent Stöfler.

HOMMEL. A huge walled plain, 75 miles in diameter, sufficiently far from the limb to be quite well seen under good libration. It has walls of fair height, and is notable because the floor has been broken by two large craters, A and C; the eastern (A) has a central mountain. Outside Hommel, to the west, are the 20-mile craters ASCLEPI and TANNERUS, while Pitiscus adjoins Hommel to the north.

JACOBI. A 41-mile crater, with walls rising in places to almost 10,000 feet. It lies south-east of the Heraclitus group. Its neighbours are KINAU (26 miles across) and LILIUS (32 miles), both of which have high walls and central peaks, but present no features of particular note. A considerable crater lies between Lilius and Jacobi.

JANSSEN. This great ruin has a diameter of over 100 miles, and is thus one of the Moon's major formations, but it is in a sad state, and is prominent only when near the terminator. Its walls are broken in the north by Fabricius and in the south by the bright-walled, 30-mile crater LOCKYER; the floor is light, and is crowded with detail, including a prominent crater-rill.

KINAU. This well-marked crater has been described with Jacobi.

LICETUS. A member of the Heraclitus group, and described with it.

LILIUS. This has already been described with Jacobi.

LINDENAU. This is described with Rabbi Levi.

MANZINUS. A large crater, 55 miles across, and with high terraced walls reaching 14,000 feet here and there. It lies in the Boguslawsky area, rather further away from the limb.

MAUROLYCUS. A walled plain with an average diameter of 68 miles, lying east of the darker-floored Stöfler and fairly obviously associated with it. The walls are of some altitude, but are broken in several places by craterlets and landslips. There is a central mountain group, and the floor in general is rough, containing some ruined rings. Immediately outside Maurolycus, to the south-east, is BAROCIUS, 50 miles across, whose high walls are broken in the north-east by two considerable craters. One of the most interesting points about the Maurolycus-Barocius group is that Maurolycus itself has encroached upon an old formation to its south; the broken formation—which also touches Barocius—is smaller than Maurolycus, so that here we do seem to have an exception to the general overlapping rule.

MILLER. This and NASIREDDIN are two 30-mile craters west of Stöfler. They make up part of the group which includes Orontius and Huggins, and which has been described in Section 11.

MORETUS. A fine walled plain 75 miles in diameter, and comparable with Theophilus and Copernicus; it is indeed larger than either, and has a high, broad, terraced wall. The floor is somewhat dark, and there is a particularly lofty central mountain crowned by a small pit. Nearby is the 29-mile, high-walled crater CYSATUS (Section 11). The mountains on the limb beyond Moretus were once thought to be the highest on the whole Moon, and were named the Leibnitz Mountains; but space-probe pictures have shown that they do not make up a true range, and the name has now been deleted from the official maps, though it has been re-allotted to a crater on the Moon's far side.

MUTUS. A 50-mile crater near Manzinus. Its walls are high, rising to peaks of 14,000 feet. Mutus has two large craters on its floor, so that it is easy to recognize; the arrangement resembles that of the distinctly larger Hommel. The area between Mutus and Tannerus, to the north, is very crowded with detail.

NASIREDDIN. This has been described with its companion Miller.

NEARCH. This 38-mile formation lies near Hommel and Hagecius, and may be regarded as a member of the Vlacq group. The floor contains several craterlets of some size, and immediately to the south is a deep crater, Nearch A, which has a central peak.

NICOLAI. A regular crater 27 miles in diameter, roughly between Janssen and Maurolycus. Nicolai has walls rising to 6,000 feet. It is right in the upland area, but there are no large craters really close to it, and this is one of the 'smoother' parts of the area, even though it is still very rough judged by general lunar standards. Well to the south-west lie various craters of little note, including SPALLANZANI and IDELER.

PENTLAND. A 45-mile crater near Curtius, with terraced walls and a double-peaked central mountain. A considerable crater, Pentland A, lies to the south. Nearby lies ZACH, which is about the same in form.

PICCOLOMINI. A splendid crater at the end of the Altai Scarp. It is 56 miles across, with high, terraced walls which include peaks rising to 15,000 feet; the southern wall has unusual structure. To the north-west, on the eastern side of the Scarp, there is a dome. Between Piccolomini and the Lindenau group lies the well-formed crater ROTHMANN, which has a peak slightly displaced from the centre of the floor.

PITISCUS. A crater 50 miles in diameter, adjoining Hommel to the north. Its walls have peaks rising to 10,000 feet, but are narrow in places; the floor contains several craters. To the west, in the general direction of Maurolycus, lie several craters, including Ideler and Spallanzani.

PONS. A 20-mile crater close to the Altai Scarp, north-west from Piccolomini. Its walls are abnormally thick. Nearby, to the west, is the irregular, 52-mile SACROBOSCO.

PONTANUS. A well-formed 28-mile crater, about midway between the Aliacensis-Werner pair and the Altai Scarp. Its broad walls are disturbed in places. There is no central peak, but a crater lies very near the centre of the floor. Closely south-east is a smaller crater, Pontanus C.

PONTÉCOULANT. This lies on the borders of the present Section and Section 15. It is too near the limb to be well seen, but it is a major walled plain with a diameter of 60 miles, and there is considerable detail inside it.

RABBI LEVI. This is one of a group lying some way south-west of Piccolomini; the other members are RICCIUS, LINDENAU, ZAGUT, CELSIUS and WILKINS. Rabbi Levi is 50 miles across, but is irregular in shape, with rather low walls and a rough floor containing numerous pits and craterlets. Celsius, separated from Rabbi Levi by a broad valley, is smaller, and has a deep craterlet some way from the centre of the floor. Riccius (50 miles in diameter) also has broken walls and a rough, pitted floor; Lindenau is 35 miles across, with higher terraced walls and a group of low mounds near its centre; Wilkins, an irregular enclosure with a mean diameter of 40 miles; Zagut, another formation of the same type, though larger (50 miles across). The whole group is complex, but in no way particularly notable.

RICCIUS. This has been described with Rabbi Levi.

ROSENBERGER. This is included in the Vlacq group. It is 50 miles in diameter, with a darkish floor and a low central peak; in the south its wall adjoins that of a smaller crater.

SCHÖMBERGER. A large crater, over 40 miles in diameter, westward along the limb from Boguslawsky. Still closer to the limb, and very hard to study from Earth, are SCOTT and AMUNDSEN, which cannot be shown in a mean libration map; Amundsen, the shallower, seems to have been damaged by the wall of Scott. Rather better placed, again not far from Schömberger, is the rather distorted but quite prominent SIMPELIUS.

STEINHEIL. This and WATT form another splendid example of a pair of overlapping craters; they lie outside the south-west wall of Janssen. Steinheil is 45 miles across, with walls which rise to 11,000 feet in the west; Watt has some ridges and delicate craterlets on its floor. The pair is similar to that of Steinheil (Section 14), but is considerably larger.

STIBORIUS. A well-marked 23-mile crater, south of Piccolomini; it has broken into a larger, older ring. The floor contains a central peak—actually slightly asymmetrically-placed. Further south lies the smaller, rather elliptical WÖHLER, which has no points of special interest.

STÖFLER. A grand enclosure, 90 miles across, with a darkish floor which makes it identifiable under any conditions of lighting. Part of the rampart has been destroyed by the intrusion of Faraday. To the north is the 40-mile, rather irregular FERNELIUS.

VLACQ. A deep, well-formed crater 56 miles in diameter, with a central hill and walls rising to 10,000 feet in places. It is a member of the group which includes Hommel, Nearch, Rosenberger, Pitiscus and Hagecius.

WERNER. The 'twin' of Aliacensis, east of the Walter chain. Werner has a diameter of 45 miles, and is extremely regular, with high terraced walls with peaks attaining almost 15,000 feet. There is also a splendid central peak.

WRIGHT. A well-formed 18-mile crater some way west of Licetus.

WILKINS. One of the Rabbi Levi group, and described with it.

ZAGUT. This also has been described with Rabbi Levi.

Section 15

AUSTRALE, MARE. The so-called Southern Sea could not be fully explored before the space-probe era, because it lies so close to the limb as seen from Earth. It extends on to the far hemisphere, and probe pictures show that it contains various craters which are lava-flooded, though not connected—a proof that the flooding came from below (though, of course, this is now no longer a matter for debate). Mare Australe is not a well-formed sea of the Crisium type, but it can be quite prominent even in spite of its unfavourable position. Further on the disk, between the Mare and the Rheita area, are various walled plains such as BRISBANE, REIMARUS, PEIRESCIUS and Vega. Naturally, the Mare Australe is best seen immediately after full moon, when it is on the terminator.

BORDA. A low-walled, 26-mile crater between Santbech (Section 16) and Reichenbach, west of Petavius. The interior contains some detail.

BRENNER. A very irregular formation abutting on Metius and Janssen.

FRAUNHOFER. A crater 30 miles in diameter, east of the Rheita Valley and south of Furnerius. Two craters break the northwest wall.

FURNERIUS. A great walled plain 80 miles across. It is a member of the great 'Eastern Chain' which includes Petavius, Vendelinus, Langrenus, the Mare Crisium and Endymion. The walls of Furnerius are somewhat broken, particularly in the north; the floor contains details such as craterlets, hills and delicate rills, as well as one prominent bright crater, Furnerius B.

HANNO. A dark-floored, 40-mile crater near the limb in the region of Pontécoulant (Section 14).

HASE. An enclosure 48 miles in diameter, closely south of Petavius. The walls are broken and the shape rather irregular; the floor contains many tiny craterlets. It is disturbed in the south by a smaller but deeper crater, Hase D. To the outer northwest of Hase it has been said that the ridges show up at sunset

Section 15

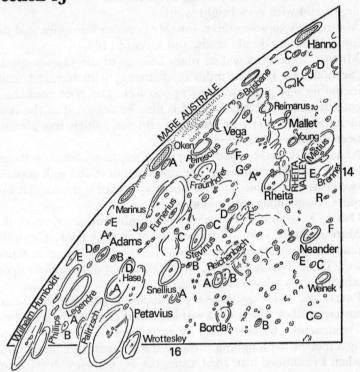

as an illuminated cross, but I have yet to see this appearance, though I have often looked for it.

HUMBOLDT, WILHELM. This formation, on the limb and close to the border of Sections 15 and 16, is 120 miles in diameter. Until photographed from the Orbiters its details were unknown, because of its extreme foreshortening; it proves to be an irregularly-outlined structure with a magnificent system of rills on its floor, but unfortunately these cannot be seen by Earth-based observers. The westward companion of Wilhelm Humboldt is PHILLIPS, 75 miles in diameter, with a long central ridge. Between Phillips and the east wall of Petavius there is a distinct crater, B.

LEGENDRE. A low-walled crater 46 miles in diameter, between Hase and Wilhelm Humboldt. The floor contains a discontinuous central ridge. South of Legendre is ADAMS, which has a central crater; closely outside it is Adams A, of considerable depth and with very bright walls.

MARINUS. A 30-mile crater, roughly between Furnerius and the limb. It has moderate walls, and a central hill.

METIUS. This large walled plain belongs to the Janssen group (Section 14). It is 50 miles in diameter, with terraced walls including one peak rising to 13,000 feet. The floor contains a considerable craterlet, B. Metius may be regarded as the twin of its south-western neighbour Fabricius, which has already been described in Section 14.

NEANDER. A well-formed crater 30 miles across, between Piccolomini and Reichenbach. The inner slopes of the wall contain two craterlets. Some distance north-westward is the smaller but quite conspicuous crater WEINEK.

OKEN. A 50-mile crater along the limb northward from the Mare Australe, easy to find because of its dark floor. The walls tend to be linear in places, but rise here and there to 6,000 feet. To the west is Oken A, which also has a darkish floor upon which stands a low central hill.

PALITZSCH. This extraordinary formation lies closely outside the eastern wall of Petavius. It was described as an irregular gorge-like structure, 60 miles long by 20 miles wide, possibly formed by a meteorite ploughing through the lunar surface layers; but when I examined it in 1952, using the 25-inch Newall refractor then at Cambridge Observatory, I found that it is nothing more nor less than a vast crater-chain, made up of several major rings which have coalesced. This has been confirmed by the Orbiter photographs. Outside, to the east, is Palitzsch A, which has continuous walls and three obvious hills upon its floor.

PETAVIUS. A magnificent crater, certainly one of the finest on the Moon. It is over 100 miles in diameter, and has walls which rise to 11,000 feet; the ramparts are very complex, and in places double. The slightly convex floor contains a grand central mountain group; the main peak exceeds 5,500 feet. A particularly conspicuous rill runs from the central area to the south-west wall. Under low and moderate illumination Petavius dominates the whole area, but it becomes obscure under very high light, and is none too easy to find at full moon. It is, of course, a member of the Eastern Chain. Palitzsch lies closely east, and abutting on Petavius to the west is the well-marked

34-mile crater WROTTESLEY, which has a twin-peaked central mountain and walls rising to 8,000 feet in places.

PHILLIPS. This has been described with its larger companion, Wilhelm Humboldt.

REICHENBACH. In itself Reichenbach is not remarkable; it is rather irregular, and about 30 miles across. North of it is Reichenbach A, which has broken into a slightly larger crater to the west of it. The chief interest in this area is the splendid REICHENBACH VALLEY, south-east of Reichenbach and narrowing steadily as it passes southward. It is really a crater-chain, and so is similar in nature to the Rheita Valley, though it is neither so conspicuous nor so well-formed.

RHEITA. A crater 42 miles across, with walls rising to 14,000 feet; the crests are unusually sharp. Associated with it is the important RHEITA VALLEY, which has been fully described in the text, and is easily seen with a small telescope when suitably lit. It is 115 miles long, and the breadth across the widest part is about 15 miles. In the Valley area there are various unremarkable rings, such as MALLET and YOUNG!

SNELLIUS. This and STEVINUS form a good example of twin formations. Each is high-walled, and about 50 miles in diameter; each has a central mountain. Snellius has rather the lighter floor. Both are easy to find except under a very high light.

STEVINUS. This is described above, with Snellius.

VEGA. A well-defined crater, 50 miles across, in the Mare Australe region. It is fairly deep, and the walls are more or less continuous; there is little detail on the floor.

WROTTESLEY. This has been described with Petavius.

Section 16

ANSGARIUS. A large ring, 50 miles in diameter, inconveniently close to the limb. The floor contains little visible detail apart from a few low hills. Slightly further on the disk, and closer to the equator, is a similar though slightly smaller crater, LA PEYROUSE, 45 miles across; there are various other lesser rings nearby.

BEHAIM. A 35-mile plain, with high walls and a central crater. The floor also includes a rill. Behaim lies between Ansgarius and Hekatæus.

Section 16

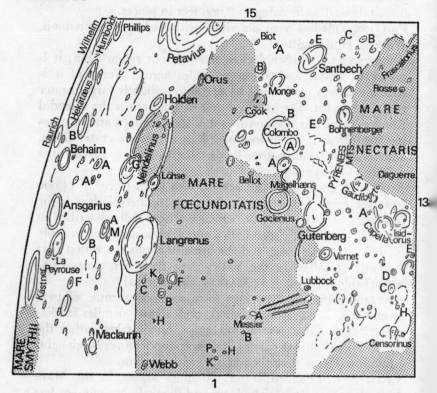

BELLOT. A 12-mile crater between Crozier and Magelhæns. It is notable because its floor is exceptionally bright.

BIOT. A 10-mile crater near the edge of the Mare Fœcunditatis, between Petavius and Santbech. A rill runs from the south wall of Biot toward the west wall of Petavius, but is a rather delicate object.

BOHNENBERGER. A rather low-walled crater, 22 miles across, on the edge of the Mare Nectaris. There is a gap in the north wall, and a ridge runs across the floor. A crater of the same size but much less well-marked, lies to the south; between this and Bohnenberger, on the rim of the old ring, is a deep craterlet. There are some very delicate rills in this area.

CAPELLA. An interesting crater in the uplands just clear of the Mare Nectaris. It is about 30 miles in diameter, and has walls which are remarkably broad in view of their moderate height. The floor contains a particularly large, rounded central mountain, crowned by a craterlet. A notable crater-valley cuts through Capella, and is traceable for a long way to either side. Capella has intruded upon a formation of similar size, ISIDORUS, lying west of it; the floor of Isidorus is relatively smooth apart from one prominent and decidedly deep craterlet.

CENSORINUS. A very small craterlet, 3 miles in diameter, in the uplands between the Mare Fœcunditatus and the southern extension of the Mare Tranquillitatis. It is one of the brightest points on the whole Moon, and is always conspicuous, particularly under high illumination. It lies on a bright patch. There is a crater, A, to the east, and a moderately large but low-walled and irregular formation to the west, touching the edge of the Mare Tranquillitatis to the north-east of Torricelli (Section 13).

COLOMBO. A 50-mile enclosure in the upland between the Mare Nectaris and the southern part of the Mare Fœcunditatis. It is disturbed in the north-west by Colombo A, about half the size and with a central hill. Well to the south-east lies COOK, which is 26 miles in diameter; the walls are low, but the crater is easy to recognize because of its dark floor. South-west of Cook is a deformed crater of similar size, MONGE, and various other unremarkable craters, including MACLURE.

CROZIER. A 15-mile crater, south-east of Colombo. It has a central hill.

FŒCUNDITATIS, MARE. Most of the Mare Fœcunditatis lies in this Section, though some of it extends on to Section 1. It is one of the less regular of the great seas, and there are not many large craters in it; the most interesting objects are probably the Messier twins. The Mare has no high mountain borders, and is connected with the Mare Tranquillitatis.

FRACASTORIUS (sometimes shortened to Fracastor). The great bay on the coast of the Mare Nectaris. It lies partly in Section 13, and has been described there.

GOCLENIUS. This makes a pair with Gutenberg. It is 32 miles in diameter, with walls rising to 5,000 feet; as it lies on the edge of the Mare Fœcunditatis, it has been flooded by the Mare lava. It has a low central hill, and the floor is cut through by a rill which extends beyond the crater. There are various other rills in the area.

GUTENBERG. Gutenberg is larger than Goclenius, since its diameter is 45 miles, but it has been distorted, and its wall is broken in the north-east by a 14-mile crater, E. There is a reduced central peak on the lava-flooded floor.

HEKATÆUS. A large, rather pear-shaped walled plain close to the limb, adjoining Wilhelm Humboldt to the north; the line is continued by Behaim, Ansgarius and La Peyrouse. Hekatæus is clearly a compound formation, and there are ridges on its floor. Several lesser rings lie near it, further on the disk.

HUMBOLDT, WILHELM. This vast enclosure lies mainly in Section 15, and has already been described. Part of Phillips is also shown in the present Section.

ISIDORUS. The companion of Capella, and described with it.

KÄSTNER. Yet another large crater too close to the limb to be well seen; it continues the chain from Wilhelm Humboldt through Hekatæus, Behaim, La Peyrouse and Ansgarius along to the Mare Smythii. Kästner is 80 miles in diameter; the floor is relatively smooth apart from a ridge running from the north wall to the centre. Between Kästner and the limb is a smaller but still considerable crater, Kästner B.

LANGRENUS. A tremendous walled plain, 85 miles across. It is a member of the Eastern Chain, and is in every way comparable with Petavius. The walls are high, massive and terraced, rising to 9,000 feet, and the floor contains a bright, twin-peaked central mountain mass. Langrenus is most imposing under low light, and appears as a bright patch near full moon. To the north-west are three craters forming a triangle: B, K and F.

LA PEYROUSE. This crater has been described with Ansgarius.

MAGELHÆNS. A 25-mile crater, south of Goclenius. It has a rather dark floor, and is joined to the south-east by a smaller crater (A) with a central hill, so that Magelhæns gives the impression of being a double formation.

MESSIER. This and its companion Messier A have already been described in the text. The curious double 'comet' ray spreads across the Mare in the general direction of LUBBOCK, an 8-mile, fairly bright craterlet; some way to the north-west of Lubbock is an irregular formation, N, with a smaller companion. Messier A used to be called 'W. H. Pickering', but this name has been officially dropped.

MONGE. An unremarkable crater, near Cook. It has been described with Colombo.

NECTARIS, MARE. Part of the Mare lies here, and part in Section 13. Craters along its eastern border include Bohnenberger and the reasonably regular, rather dark-floored GAUDIBERT. Note the bright, deep craterlet ROSSE, well on the Mare; it is 10 miles in diameter, and lies on a bright ray. Low ridges connect it with Fracastorius. Also on the Mare, west of Gaudibert, is the very low-walled DAGUERRE.

PYRENEES MOUNTAINS. Not a true range, but a collection of moderate hills, lying roughly between Gutenberg and Bohnenberger.

ROSSE. The bright craterlet on the Mare Nectaris, and described with the Mare.

SANTBECH. A 44-mile walled plain; some parts of the walls are high, and the floor is rather darkish. A depression cuts through it. Santbech lies in the uplands between the Mare Nectaris and the Mare Fœcunditatis, almost due east of Fracastorius.

VENDELINUS. A majestic irregular formation, over 100 miles from north to south. It has been considerably ruined, and seems therefore to be older than its companions in the Eastern Chain, Langrenus and Petavius, which are of much the same size. The floor of Vendelinus is darkish, and there is no central peak. To the north-east the rampart is broken by the intrusion of a 45-mile crater which certainly merits a separate name (on some previous maps it was called Smith); the rather smaller LOHSE, with a central peak, intrudes on the north-west; and to the south of Vendelinus lies HOLDEN, 25 miles in diameter and decidedly deep. West of Holden is another crater, B, deserving of a separate name. Vendelinus and its companions make a grand picture under oblique illumination.

WEBB. A bright 14-mile crater, with a darkish floor and a central hill; it is the centre of a very short, inconspicuous system of bright rays. It lies near the border of the Mare Fœcunditatis, close to the lunar equator. In the uplands to the east is MACLAURIN, which has a diameter of 30 miles; the walls are uneven, and the floor is distinctly concave. The MARE SMYTHII lies on the limb in this region, but most of it is in Section 1, and has been described there.

Appendix VII

INDEX TO FORMATIONS
DESCRIBED IN THE MAP AND TEXT

(Roman = general text; Italic = entry in Map Section; Bold = Map Sec.)

Abenezra, *286*, **13**
Abulfeda, *287*, **113**
Acherusia, Cape, *250*, **4**
Adams, *299*, **15**
Ænarium, Cape, *285*, **12**
Æstuum, Sinus, *252*, **5**
Agarum, Cape, *236*, **1**
Agatharchides, *267*, **9**
Agrippa, *245*, **4**
Airy, *287*, **13**
Albategnius, *287*, **13**
Alexander, *242*, **3**
Alfraganus, *287*, **13**
Alhazen, 199, *235*, **1**
Aliacensis, 116, 186, *291*, **14**
Almanon, *287*, **13**
Alpetragius, 199, *280*, **12**
Alphonsus, 109, 113–4, 154, 185, 203–5, 222, *280*, **12**
Alpine Valley, 102, 105, 176, 186, *241*, **3**
Alps, 102–3, 108, *240*, **3**
Altai Scarp, 103, *287*, *291*, **13, 14**
Ampère, *253*, **5**
Amundsen, *297*, **14**
Anaxagoras, *255*, **6**
Anaximander, *255*, **6**
Anaximenes, *258*, **6**
Ándĕl, *287*, **13**
Ångström, *263*, **8**
Ansgarius, *301*, **16**
Apennines, 102, 103, 164, 192, *245*, *252*, **4–5**

Apianus, *291*, **14**
Apollonius, *235*, **1**
Arago, *246*, **4**
Aratus, *247*, **4**
Archimedes, 114, 155, 183, *253*, **5**
Archytas, *241*, **3**
Argæus, *247*, **4**
Argelander, *287*, **13**
Ariadæus, 108–9, *247*, **4**
Aristarchus, 108, 117, 131, 206–7, 210, 222, *261–2*, **8**
Aristillus, *247*, **4**
Aristoteles, *242*, **3**
Arnold, *243*, **3**
Arzachel, 112–4, 203–4, *280*, **12**
Asclepi, *294*, **14**
Atlas, *238*, **2**
Australe, Mare, 113, 182, 183, 192, *298*, **15**
Autolycus, 114, *247*, **4**
Auwers, *248*, **4**
Azophi, *288*, **13**
Azout, *236*, **1**

Babbage, *259*, **6**
Baco, *293*, **14**
Baillaud, *244*, **3**
Bailly, 113, 183, *273*, **11**
Baily, *242*, **3**
Ball, *276*, **11**
Barocius, *294*, **14**
Barrow, *242*, **3**
Bayer, *279*, **11**
Beaumont, *288*, **13**
Beer, 186, *253*, **5**
Behaim, *301*, **16**
Bellot, *302*, **16**

Bernouilli, *238*, **2**
Berosus, *238*, **2**
Berzelius, *239*, **2**
Bessarion, *262*, **8**
Bessel, 188, 199, *247*, **4**
Bettinus, *274*, **11**
Bianchini, *256*, **6**
Biela, *291*, **14**
Billy, 101, 113, *267*, **9**
Biot, *302*, **16**
Birmingham, *256*, **6**
Birt, *281*, **12**
Blagg, *283*, **12**
Blanc, Mont, *241*, **3**
Blancanus, *274*, **11**
Blanchinus, *291*, **14**
Bode, *253*, **5**
Boguslawsky, *291*, **14**
Bohnenberger, *302*, **16**
Bond, G. P., *240*, **2**
Bond, W. C., *242*, **3**
Bonpland, *281*, **12**
Borda, *298*, **15**
Boscovich, *247*, **4**
Bouguer, *257*, **6**
Boussingault, *291*, **14**
Bouvard, *272*, **10**
Bradley, Mount, *245*, **4**
Brayley, *263*, **8**
Breislak, *293*, **14**
Brenner, *298*, **15**
Briggs, *263*, **8**
Brisbane, *298*, **15**
Brown, *279*, **11**
Bruce, *283*, **12**
Buch, *292*, **14**
Bulliardus, *281–2*, **12**
Burckhardt, *240*, **2**
Bürg, 107, *242*, **3**
Burnham, *288*, *291*, **13**

Büsching, *292*, **14**
Byrgius, *270*, **10**

Cabæus, *273*, **11**
Calippus, *242*, **3**
Campanus, *275*, **11**
Capella, *302*, **16**
Capuanus, 106, *275*, **11**
Cardanus, *263*, **8**
Carlini, *253*, **5**
Carpathian Mts., 103, *253*, **5**
Carpenter, *255*, **6**
Carrington, *239*, **2**
Casatus, *275*, **11**
Cassini, *242*, **3**
Cassini, J. J., *257*, **6**
Catharina, 114, *288*, **13**
Caucasus Mts., 103, 108, 192, *242*, **3**
Cauchy, *236*, **1**
Cavalerius, *263*, **8**
Cavendish, *270*, **10**
Cayley, *247*, **4**
Celsius, *296*, **14**
Censorinus, *303*, **16**
Cepheus, *238*, *239*, **9**
Chacornac, *247*, **4**
Challis, *242*, **3**
Chevallier, *238*, **2**
Chladni, *250*, **4**
Cichus, *275*, **11**
Clairaut, *293*, **14**
Clausius, *273*, **10**
Clavius, 112, 113, 158, 183, 186, 190, *275*, **11**
Cleomedes, 131, 187, 199, *236*, **1**
Cleostratus, *259*, **7**
Colombo, *303*, **16**
Condamine, *257*, **6**
Condorcet, *236*, **1**
Conon, *245*, **4**
Cook, *303*, **16**
Copernicus, 81, 114–5, 119, 142, 160, *253*, **5**
Cordillera Mts., *267*, *272*, **9**

Crisium, Mare, 55, 96, 97, 100, 158, 159, 181, 182, 192, 208, *236*, **1**
Crozier, *303*, **16**
Crüger, 101, 113, *267–8*, **9**
Curtius, *293*, **14**
Cusanus, *244*, **3**
Cuvier, *293*, **14**
Cyrillus, 114, 187, *288*, **13**
Cysatus, *276*, *295*, **11**

Daguerre, *305*, **16**
D'Alembert Mts., 266, *268*, **9**
Damoiseau, *268*, **9**
Daniell, *245*, **3**
Darney, *283*, **12**
D'Arrest, *248*, **4**
Darwin, 106, *268*, **9**
Davy, *282*, **12**
Dawes, *247*, **4**
Debes, *236*, **1**
Dechen, *259*, **7**
De Gasparis, *270*, **10**
Delambre, *288*, **13**
de la Rue, *243*, **3**
Delaunay, *288*, **13**
De l'Isle, *263*, **8**
Delmotte, *236*, **1**
Deluc, *276*, **11**
Dembowski, *248*, **4**
Democritus, *243*, **3**
Demonax, *293*, **14**
De Morgan, *247*, **4**
Descartes, 164, 165, *287*, **13**
Deslandres, *276*, **11**
Deseilligny, *247*, **4**
De Vico, *268*, **9**
Dionysius, 248, **4**
Diophantus, *263*, **8**
Dollond, *287*, **13**
Donati, *288*, **13**
Doppelmayer, 99, 191, *270*, **10**

Dörfel Mts., *273*, **11**
Draper, *254*, **5**
Drebbel, *272*, **10**
Drygalski, *273*, **11**
Dunthorne, *275*, **11**

Egede, *243*, **3**
Eichstädt, *271*, **10**
Eimmart, *236*, **1**
Einstein, 97, 151–2, *265–6*, **8**
Elger, *275*, **11**
Encke, *263*, **8**
Endymion, 131, 183, *239*, **2**
Epidemiarum, Palus, *276*, **11**
Epigenes, *257*, **6**
Epimenides, *276*, **11**
Eratosthenes, 102, 116, 133, *253*, **5**
Euclides, 103, 119, *268*, **9**
Euctemon, *244*, **3**
Eudoxus, *243*, **3**
Euler, *254*, **5**

Fabricius, *293*, **14**
Faraday, *293*, **14**
Fauth, *254*, **5**
Faye, *288*, **13**
Fermat, *287*, **13**
Fernelius, *297*, **14**
Feuillé, 186, *253*, **5**
Firmicus, *236*, **1**
Flammarion, *282*, **12**
Flamsteed, *268*, **9**
Fœcunditatis, Mare, 161, *303*
Fontana, *268*, **9**
Fontenelle, 200, *257*, **6**
Foucault, *257*, **6**
Fourier, *271*, **10**
Fracastorius, 115, 289, *303*, **16**
Fra Mauro, 163, 210, *282*, **12**
Franklin, *238*, **2**

Franz, *237*, **1**
Fraunhofer, *298*, **15**
Fresnel, Cape, *245*, **4**
Frigoris, Mare, 182–3, 192, *243*, *257*, **3**, **6**
Furnerius, 298

Galileo, *263*, **8**
Galle, *243*, **3**
Galvani, *259*, **7**
Gambart, *283*, **12**
Gärtner, *243*, **3**
Gassendi, 109, 207, 222, *268*, **9**
Gaudibert, *305*, **16**
Gauricus, *276*, **11**
Gauss, *240*, **2**
Gay-Lussac, *254*, **5**
Geber, *289*, **13**
Geminus, *240*, **2**
Gemma Frisius, *293*, **14**
Gerard, 149–50, *259*, **7**
Gioja, *243*, **3**
Goclenius, *303*, **16**
Godin, *248*, **4**
Goldschmidt, *243*, **3**
Goodacre, *293*, **14**
Gould, *283*, **12**
Grimaldi, 55, 81, 101, 113, 142, 155, 159, 183, 191, 210, 222, *269*, **9**
Grove, *244*, **3**
Gruemberger, *276*, **11**
Gruithuisen, *263*, **8**
Guericke, 154, *282*, **12**
Gutenberg, *304*, **16**
Gyldén, *283*, **12**

Hadley, Mount, 102, *245*, **4**
Hæmus Mts, 103, *248*, **4**
Hagecius, *293*, **14**
Hahn, *238*, **2**
Hainzel, *276*, **11**
Hall, *240*, **2**
Halley, *289*, **12**
Hanno, *298*, **15**

Hansen, *235*, **1**
Hansteen, *267*, *269*, **9**
Harbinger Mts., 102, 103, *264*, **8**
Harding, *259*, **7**
Harpalus, *257*, **6**
Hase, *298*, **15**
Hausen, *273*, **11**
Heinsius, *276*, **11**
Heis, *254*, **5**
Hekatæus, *304*, **16**
Helicon, *257*, **6**
Hell, 276, *277*, **11**
Helmholtz, *292*, **14**
Heraclides, Cape, *258*, **6**
Heraclitus, *293*, **14**
Hercules, *244*, **3**
Hercynian Mts., *264*, **8**
Herigonius, *269*, **9**
Hermann, *269*, **9**
Herodotus, 108, 206, *264*, **8**
Herschel, *282*, **12**
Herschel, C., *259*, **7**
Herschel, J., *257*, **6**
Heriodus, *277*, **11**
Hevel, *264*, **8**
Hind, *289*, **13**
Hippalus, 116, 191, *272*, **10**
Hipparchus, *289*, **13**
Holden, *305*, **16**
Hommel, *294*, **14**
Hooke, *240*, **2**
Horrebow, *257*, **6**
Horrocks, *289*, **13**
Hortensius, *264*, **8**
Huggins, *278*, **11**
Humboldt, Wilhelm, 94, 158, *299*, *304*, **15**
Humboldtianum Mare, 159, *240*, **2**
Humorum, Mare, 99, 101, 107, 159, 182, *269*, 271–2, **9**, **10**
Huygens, Mount, 102, *252*, **5**

Hyginus, 108–9, 117, 183, 185, 200, *248*, **4**
Hypatia, *287*, **13**

Ideler, *296*, **14**
Imbrium, Mare, 74, 99, 100, 102–3, 105, 182–3, 185, 191, 192, 208, 221, *254*, *257*, **5**, **6**
Inghirami, *272*, **10**
Iridum Sinus, 102, 116, 159, 186, *258*, *260*, **6**, **7**
Isidorus, *304*, **16**

Jacobi, *294*, **14**
Jansen, *250*, **4**
Janssen, 113, *294*, **14**
Julius Cæsar, *248*, **4**
Jura Mts., 102, 103, *258*, **6**

Kaiser, *291*, **14**
Kane, *243*, **3**
Kant, *289*, **13**
Kästner, *304*, **16**
Kelvin, Cape, *272*, **10**
Kepler, 119, *264*, **8**
Kies, *277*, **11**
Kinau, *294*, **14**
Kirch, *258*, **6**
Kircher, *274*, **11**
Kirchhoff, *237*, **1**
Klaproth, *277*, **11**
Klein, *287*, **13**
König, *277*, **12**
Krafft, *263*, **8**
Krieger, *263*, **8**
Krusenstern, *291*, **14**
Kunowsky, *264*, **8**

Lacaille, *278*, **11**
Lacroix, *273*, **10**
Lade, *289*, **13**
Lagalla, *277*, **11**
Lagrange, *272*, **10**
La Hire, *254*, **5**

Lalande, *283*, **12**
Lambert, *254*, **5**
Lamèch, *243*, **3**
Lamont, *246*, **4**
Landsberg, *283*, **12**
Langrenus, 183, *304*, **16**
La Peyrouse, *301*, *304*, **16**
Laplace, Cape, *258*, **6**
Lassell, *282*, **12**
Lavinium, Cape, *237*, **1**
La Voisier, *259*, **7**
Lee, *271*, **11**
Legendre, *299*, **15**
Legentil, *273*, **11**
Lehmann, *272*, **10**
Le Monnier, 99, 115, 161, 191, *248*, **4**
Lepaute, *276*, **11**
Letronne, *269*, **9**
Le Verrier, 257, *258*, **6**
Lexell, *276*, **11**
Licetus, *293*, **14**
Lichtenberg, *260*, **7**
Lick, *237*, **1**
Liebig, *270*, **10**
Lilius, *294*, **14**
Lindenau, *294*, *296*, **14**
Linné, 88, 142, 197–9, *248*, **4**
Lippershey, *277*, **11**
Littrow, *248*, **4**
Lockyer, *294*, **14**
Loewy, *272*, **10**
Lohrmann, *269*, **9**
Lohse, *305*, **16**
Longomontanus, *277*, **11**
Louville, *261*, **7**
Lubbock, *304*, **16**
Lubiniezky, *283*, **12**
Luther, *249*, **4**
Lyell, *237*, **1**

Maclaurin, *305*, **16**
Maclear, *249*, **4**
Maclure, *303*, **16**
Macrobius, *236*, **1**

Mädler, *289*, **13**
Magelhæns, *304*, **16**
Maginus, *277*, **11**
Main, *242*, **3**
Mairan, *260*, **7**
Malapert, *273*, **11**
Mallet, *301*, **15**
Manilius, *248*, **4**
Manners, *246*, **4**
Manzinus, *294*, **14**
Maraldi, *249*, *251*, **4**
Marco Polo, *253*, **5**
Marginis, Mare, *236*, **1**
Marinus, *300*, **15**
Marius, *264*, **8**
Marth, *278*, **11**
Maskelyne, *249*, **4**
Mason, *244*, **3**
Maupertuis, *256*, **6**
Maurolycus, *294*, **14**
Maury, *238*, **2**
Mayer, C., *244*, **3**
Mayer, T., *265*, **8**
Medii, Sinus, 156, *283*, *290*, **12**, **13**
Mee, *276*, **11**
Menelaus, *249*, **4**
Mercator, *277*, **11**
Mercurius, *240*, **2**
Mersenius, *269*, **9**
Messala, *240*, **2**
Messier, 186, 200, *304*, **16**
Metius, *300*, **15**
Meton, *244*, **3**
Milichius, 106, *265*, **8**
Miller, *294*, **14**
Mitchell, *242*, **3**
'Miyamori Valley', *269*, **9**
Moigno, *243*, **3**
Möltke, *291*, **13**
Monge, *305*, **16**
Montanati, *277*, **11**
Moretus, *295*, **14**
Mortis, Lacus, *244*, **3**
Mösting, *283*, **12**
Möstlin, *263*, **8**

Müller, *289*, **13**
Murchison, *254*, **5**
Mutus, *295*, **14**

Nasireddin, *295*, **14**
Nasmyth, *278*, **11**
Naumann, *259*, **7**
Neander, *300*, **15**
Nearch, *295*, **14**
Nebularum, Palus, *244*, *258*, **3**, **6**
Nectaris, Mare, 101, 159, 182, 191, 290, *305*, **13**, **16**
Neison, *244*, **3**
Neper, *236*, **1**
Neumayer, *292*, **14**
Newcomb, *236*, **1**
Newton, 111, *278*, **11**
Nicolai, *295*, **14**
Nicollet, *283*, **2**
Nöggerath, *278*, **11**
Nonius, *291*, **14**
Nubium, Mare, 99, 154, 171, 182, 278, *283–4*, **11**, **12**

Œnopides, *261*, **7**
Œrsted, *238*, **2**
Oken, *300*, **15**
Olbers, 119, *265*, **8**
Olivium, Cape, *237*, **1**
Opelt, *283*, **12**
Oppolzer, *284*, **12**
Oriani, *237*, **1**
Orientale, Mare, 10, 93, 101, 152, 157, 159, 168, 182, 184, 191, 192, 232, 266, 267, 272, 312, **10**
Orontius, *278*, **11**

Palisa, *282*, **12**
Palitzsch, *300*, **15**
Pallas, *264*, **5**
Palmieri, *271*, **10**
Parrot, *290*, **13**
Parry, *284*, **12**

Peirce, *236, 237*, **1**
Peirescius, *298*, **15**
Pentland, *296*, **14**
Petavius, *300*, **15**
Petermann, *244*, **3**
Peters, *244*, **3**
Phillips, *299, 301*, **15**
Philolaus, *258*, **6**
Phocylides, *278*, **11**
Piazzi, *272*, **10**
Piazzi Smyth, *258*, **6**
Picard, *236, 237*, **1**
Piccolomini, 103, *296*, **14**
Pickering, E. C., *289*, **13**
Pico, 103, *258*, **6**
Pictet, *279*, **11**
Pietrosul Bay, *254*, **5**
Pingré, *278*, **11**
Pitatus, *278*, **11**
Pitiscus, *296*, **14**
Piton, 104, *258*, **6**
Plana, *244*, **3**
Plato, 81, 102, 113, 142, 183, 201, 203, 258–9, **6**
Playfair, *286*, **13, 14**
Plinius, *249*, **4**
Plutarch, *237*, **1**
Poisson, *291*, **14**
Polybius, *290*, **13**
Pons, *296*, **14**
Pontanus, *296*, **14**
Pontécoulant, *296*, **14**
Porter, *275*, **11**
Posidonius, *249*, **4**
Prinz, 106, *263, 265*, **8**
Procellarum, Oceanus, 101, 106, 155–6, 191, 192, *261, 265*, 270, **7, 8, 9**
Proclus, *237*, **1**
Proctor, *277*, **11**
Protagoras, *242*, **3**
Ptolemæus, 55, 111, 112–4, 183, 203, *284*, **12**

Puiseux, *272*, **10**
Purbach, *278*, **11**
Putredinis, Palus, *249*, 255, **4, 5**
Pyrenees Mts., *305*, **16**
Pythagoras, 112, 183, 259, **5**
Pytheas, 255, **5**

Rabbi Levi, *296*, **14**
Ramsden, *276*, **11**
Réaumur, *284, 290*, **12, 13**
Regiomontanus, *279*, **11**
Régnault, *259*, **7**
Reichenbach, *301*, **15**
Reimarus, *298*, **15**
Reiner, *264, 265*, **8**
Reinhold, 255, **5**
Repsold, *261*, **7**
Rhæticus, *290*, **13**
Rheita, *301*, **15**
Rheita Valley, 105, 183, 185, *301*, **15**
Riccioli, 81, 113, 270, **9**
Riccius, 296, 297, **14**
Riphæan Mts., 103, *284*, **12**
Ritchey, *289*, **13**
Ritter, *249*, **4**
Robinson, *257, 259*, **6**
Rocca, *270*, **9**
Rømer, *237*, **1**
Rook Mts., *267, 272*, **10**
Roris, Sinus, *261*, **7**
Rosenberger, *297*, **14**
Ross, *249*, **4**
Rosse, *305*, **16**
Rost, 280, **11**
Rothmann, *296*, **14**
Rümker, *261*, **7**
Rutherford, *275*, **11**

Sabine, *250*, **4**
Sacrobosco, *296*, **14**
Santbech, *305*, **16**
Sasserides, *279*, **11**

Saunder, *289*, **13**
Saussure, *278*, **11**
Scheiner, *274, 279*, **11**
Schiaparelli, 265, **8**
Schickard, 183, *272*, **10**
Schiller, *279*, **11**
Schmidt, *249*, **4**
Schneckenberg, Mount, *248, 250*, **4**
Schömberger, *297*, **14**
Schröter, *284*, **12**
Schröter's Valley, 108, 284
Schubert, *237*, **1**
Schumacher, *240*, **2**
Schwabe, *244*, **3**
Scoresby, *244*, **3**
Scott, *297*, **14**
Secchi, *238*, **1**
Seeliger, *290*, **13**
Segner, *274*, **11**
Seleucus, *265*, **8**
Seneca, *237*, **1**
Serao, *253*, **5**
Serenitatis, Mare, 99, 100, 103, 159, 182, 192, 208, *250*, **4**
Sharp, *261*, **7**
Sheepshanks, *244*, **3**
Short, *295*, **14**
Shuckburgh, *238*, **2**
Silberschlag, *250*, **4**
Simpelius, *297*, **14**
Sinas, *250*, **4**
Sirsalis, 186, *270*, **9**
Sirsalis Rill, *270*, **9**
Smythii, Mare, 159, *237, 305*, **1, 16**
Snellius, *301*, **15**
Sömmering, *284*, **12**
Somnii, Palus, *237*, **1**
Somniorum, Lacus, *244*, **3**
Sosigenes, *250*, **4**
South, *259*, **6**
Spallanzani, *296*, **14**
Spitzbergen Mts., *259*, **5, 6**

Spörer, *282*, **12**
Spumans, Mare, *235*, **1**
Stadius, 116, 192, *255*, **5**
Stag's-Horn Mts., 107, *284*, **12**
Steinheil, 116, 186, *297*, **14**
Stevinus, *301*, **15**
Stiborius, *297*, **14**
Stöfler, *297*, **14**
Strabo, *245*, **3**
Straight Range, 103, *259*, **6**
Straight Wall, 107, 176, *284*, **12**
Street, *279*, **11**
Struve, *240*, **2**
Struve, Otto, 151, *265*, **8**
Suess, *265*, **8**
Sulpicius Gallus, *250*, **4**
Sven Hedin, *266*, **8**

Tacitus, *290*, **13**
Tannerus, *294*, **14**
Taquet, *250*, **4**
Taruntius, 114, 186, *238*, **1**
Taurus Mts., *237*, *240*, **1, 2**
Taylor, *287*, **13**
Tempel, *247*, **4**
Teneriffe Mts., *258–9*, **6**
Thales, 202, *245*, **3**
Theætetus, 111, 203, *245*, **3**
Thebit, 107, 186–7, *285*, **12**

Theon Junior, *290*, **13**
Theon Senior, *290*, **13**
Theophilus, 110, 112, 114, 187, *290*, **13**
Timæus, *245*, **3**
Timocharis, *255*, **5**
Timoleon, *240*, **2**
Tisserand, *236*, **1**
Torricelli, *290*, **13**
Tralles, 187, *236*, **1**
Tranquillitatis, Mare, 154, 156, *238*, *250*, *291*, **1, 4, 13**
Triesnecker, 109, *250*, **4**
Trouvelot, *241*, **3**
Turner, *285*, **12**
Tycho, 81, 118, 142, 176, 185, 187, 188, *279*, **11**

Ukert, *251*, **4**
Ulugh Beigh, *260*, **7**
Undarum, Mare, *238*, **1**
Ural Mts., *284*, **12**

Vaporum, Mare, 102, *251*, **4**
Vasco da Gama, *266*, **8**
Vega, 298, *301*, **15**
Vendelinus, 183, *305*, **16**
Vieta, *273*, **10**
Virgil, *262*, **8**
Vitello, 114, 186, *273*, **10**
Vitruvius, *251*, **4**
Vlacq, 183, *297*, **14**
Vogel, 185, *291*, **13**

Wallace, *255*, **5**
Walter, 183, *279*, **11**
Wargentin, 117, 191, *273*, **10**
Watt, 116, *297*, **14**
Webb, *305*, **16**
Weigel, *280*, **11**
Weinek, *300*, **15**
Weiss, *275*, **11**
Werner, 86, 186, *297*, **14**
Whewell, *247*, **4**
Wichmann, *270*, **9**
Wilhelm I, *279*, **11**
Wilkins, *296*, *298*, **14**
Williams, *244*, **3**
Wilson, *274*, **11**
Wöhler, *297*, **14**
Wolf, Max, *283*, **12**
Wolff, Mount, *253*, **5**
Wollaston, *263*, **8**
Wright, *297*, **14**
Wrottesley, *301*, **15**
Wurzelbauer, *280*, **11**

Xenophanes, *261*, **7**

Yerkes, *237*, **1**
Young, *301*, **15**

Zach, *296*, **14**
Zagut, *296*, *298*, **14**
Zeno, *240*, **2**
Zöllner, *287*, **13**
Zucchius, *274*, *280*, **11**
Zupus, *270*, **9**

LATIN AND ENGLISH NAMES
OF THE LUNAR SEAS

Mare Australe	The Southern Sea
Mare Crisium	The Sea of Crises
Palus Epidemiarum	The Marsh of Epidemics
Mare Fœcunditatis	The Sea of Fertility
Mare Frigoris	The Sea of Cold
Mare Humboldtianum	Humboldt's Sea
Mare Humorum	The Sea of Humours
Mare Imbrium	The Sea of Showers
Sinus Iridum	The Bay of Rainbows
Mare Marginis	The Marginal Sea
Sinus Medii	The Central Bay
Lacus Mortis	The Lake of Death
Palus Nebularum	The Marsh of Mists
Mare Nectaris	The Sea of Nectar
Mare Nubium	The Sea of Clouds
Mare Orientale	The Eastern Sea
Oceanus Procellarum	The Ocean of Storms
Palus Putredinis	The Marsh of Decay
Sinus Roris	The Bay of Dews
Mare Serenitatis	The Sea of Serenity
Sinus Æstuum	The Bay of Heats
Mare Smythii	Smyth's Sea
Palus Somnii	The Marsh of Sleep
Lacus Somniorum	The Lake of the Dreamers
Mare Spumans	The Foaming Sea
Mare Tranquillitatis	The Sea of Tranquillity
Mare Undarum	The Sea of Waves

Appendix VIII

THE FAR SIDE OF THE MOON

TO MAKE THIS BOOK reasonably complete, it seems only right to include a map of the far side; but there is no point in giving too much detail, because there are so few people who will ever be able to see the features direct! For the same reason, I have followed the official practice of putting north at the top.

There are obvious differences between the far and the near hemispheres. Apart from the Mare Orientale, which spans both, the far side has no major maria, but there are huge 'unfilled' basins such as Hertzsprung and Korolev. The general aspect is of upland, and the laws of distribution apply. There is also a certain uniformity in the distribution of the formations; I have already mentioned the great arc which includes Birkhoff, D'Alembert (not to be confused with the D'Alembert Mountains on the near hemisphere—a name not included in the latest IAU list), Campbell, the Mare Moscoviense, Mendeleev, Gagarin and Mare Ingenii, as well as various others. Among special features mention should be made of the tremendous valleys of Schrödinger and Planck, in the north, not very far beyond the Earth-turned limb at maximum libration.

This, then, is a mere outline chart, and I have named only the leading formations. The nomenclature follows that approved by the IAU. The names range from those of famous novelists (Jules Verne, H. G. Wells) through to modern astronauts.

MAP OF THE FAR SIDE OF THE MOON

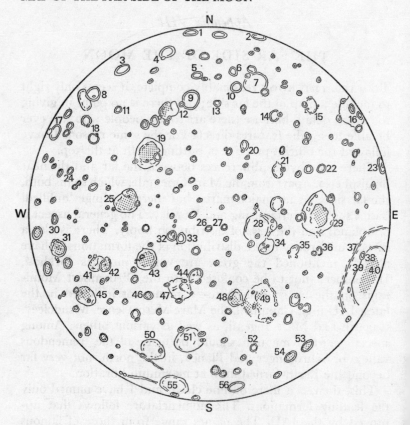

Key:

1	Nansen	17	Joliot
2	Brianchon	18	Szilard
3	Compton	19	MARE MOSCOVIENSE
4	Schwarzschild	20	Cockcroft
5	Avogadro	21	Mach
6	Sommerfeld	22	Fersman
7	Birkhoff	23	Einstein
8	Fabry	24	Fleming
9	D'Alembert	25	Mendeléev
10	Rowland	26	Dædalus
11	Wells	27	Icarus
12	Campbell	28	Korolev
13	Dunér	29	Hertzsprung
14	Fowler	30	Hirayama
15	Landau	31	Pasteur
16	Lorentz	32	Keeler

33 Heaviside
34 Galois
35 Paschen
36 Ioffe
37 CORDILLERA MOUNTAINS
38 Lowell
39 ROOK MOUNTAINS
40 MARE ORIENTALE
41 Fermi
42 Tsiolkovskii
43 Gagarin
44 Van de Graaff

45 Milne
46 Jules Verne
47 MARE INGENII
48 Oppenheimer
49 Apollo
50 Planck
51 Poincaré
52 Minkowski
53 Mendel
54 Boltzmann
55 Schrödinger
56 Zeeman

GENERAL INDEX

(See also separate index of Formations on the Moon, pp. 306-11)

Adamski, G., 132-3
Agung, Mount, 140
Airy, Sir G., 122
Aldrin, E., 11, 134, 146, 163, 164, 206, 210, 285
Allen, M., 64
Alter, D., 203-4
Anaxagoras, 138
Anaximander, 17
Anders, W., 94, 163
Anslow, L., 122
Antoniadi scale, 221
Apollo (lunar crater), 184
Apollo-8, 94, 163
   -9, 163
   -10, 163
   -11, 11, 163, 206, 285
   -12, 163, 171, 266, 284
   -13, 162
   -14, 162, 169, 282
   -15, 102, 104, 109, 170-1, 206
   -16, 285, 287
   -17, 125, 161, 168
Appleton, 63
Arago, F., 90
Aristarchus, 18
Armacolite, 171
Armstrong, N., 11-12, 134, 146, 163, 166, 206, 210, 285
Arthur, D. W. G., 109
Ashbrook, J., 107
Asteroids, *see* Minor Planets
Asthenosphere, lunar, 165
Astrology, 70
Atlantis, 174
Auzout, A., 79

B.A.A. (British Astronomical Association), 89, 203, 222, 224
B.E.M.s, 126-7
Baldwin, R. B., 109, 178-9, 183
Barker, R., 106, 236
Barnard, E. E., 123, 202
Barr, E., 206
Barringer Crater, *see* Meteor Crater
Barwell Meteorite, 27
Barycentre, the, 49-50
Basalts, 164
Bean, A., 168
Beard, D. P., 176
Beer, W., 85, 88, 129, 139, 197-8, 200
Berkeley, G., 68
Birkoff (lunar crater), 184
Birmingham, J., 82

Birt, W. R., 89
Bjorkholm, P., 207
Blagg, M. A., 91
Blowhole craters, 117
Blue Moons, 66
Boneff, N., 64, 177
Boring, E. G., 68
Borman, F., 94, 163
Boyce, P., 206
Breccias, 164, 171
Brinton, H., 121
Bruno, G., 19

Callisto, 25
Campbell (lunar crater), 184, 312
Carpenter, J., 189-90
Cassini, G. D., 82, 199
Ceres, 23
Cernan, E., 125, 164, 169, 213
Chapman, D., 71
Charbonneaux, 203, 245
Chubb Crater, 27, 180
Collins, M., 11-12, 134, 163, 206
Columbus, Christopher, 139
Conrad, C., 168, 171
Comets, 25-6
Continental Drift, 34
Cooke, S. R. B., 106
Copernican Revolution, 19
Copernicus, N., 19, 48
Crab Nebula, 123
Cragg, T. A., 203
Craterlets, 177
Craters, lunar, 110-19
   chains of, 184
   classification of, 110
   depth/diameter ratios of, 81
   distribution of, 113, 116, 183-7, 192
   forms of, 111, 181
   profiles of, 112
   relative ages of, 113, 192
   slope angles of, 111, 190
   summit pits, 188
Craters, origin of, 173-93
   atomic bomb theory, 175-6
   bubble theory, 190
   caldera theory, 190-1
   contraction theory, 177-8
   coral atoll theory, 176
   crystallization theory, 178
   expansion theory, 178
   fluidized bed theory, 192-3
   ice theory, 174-5

impact theory, 178–89
tidal theory, 177
volcanic foundation theory, 189–90
Craters, Martian, 193–4
Mercurian, 195–6
terrestrial impact, 27–8, 179–81
Venusian, 195
Cyrano de Bergerac, 145

D'Alembert (lunar crater), 184, 312
Daguerreotypes, 90
Danjon, A., 140
Scale, 140
Darwin, Sir G. H., 32–4
Davison, C., 64
de la Rue, W., 90
Deimos, 22–3, 40, 54, 194
Dollfus, A., 124
Domes, lunar, 106
Douglass, A. E., 123, 142
Draper, J. W., 90
Drawings, lunar, 220–1

Eagle, 11–12
Earth, age of, 31
second satellite of?, 41
size of, 18
status of, 16–17
Earthquakes, 64–5
Earthshine, 16
Eclipses, lunar, 18, 50, 52, 135–43,
228–9
causes of, 136
colours during, 139
Danjon scale of, 140
effects on lunar surface, 141–2
historical, 138–9
list of, 228
penumbral, 137
Eclipses, solar, 51, 135
Elger, T. G., 89
Emley, E. F., 151
Encke's Comet, 26
Engel, K. H., 178
Eratosthenes, 18
Ericson, 174
Etna, Mount, 191
Europa, 24

Faults, lunar, 106–7
Fauth, P., 175
Fermi (lunar crater), 158
Fesenkov, V., 124
Fielder, G., 92, 107, 108, 192
Fillias, A., 178
Firsoff, V. A., 107, 222
Fission theory, 32–4
Fitton, L. E., 123
Flashes on the Moon, 201–2

Fluorescence, lunar?, 208
Flying Saucers, 132
Foreshortening of lunar formations, 94,
96, 112
Franz, J. H., 91
Furlong, T. J., 68–9

Gagarin (lunar crater), 184, 312
Galaxy, the, 29
Galileo, 20, 75–9, 104, 157, 263
Ganymede, 25
Geake, J., 208
Gegenschein, the, 28
Gilbert, G. K., 178, 180
Gilbert, W., 74–5
Glows, lunar, 202
Goddard, R. H., 146
Godwin, F., 144
Gold, T., 154, 179
dust theory by, 154–5, 164
flash theory by, 165
Goodacre, W., 91
Gorenstein, P., 207
Great Rift Valley, 191
Greeks, the, 17–18, 45, 47
Green, J., 191, 209
Greenacre, J., 206
Gregory, R., 68–9
Grid system, lunar, 107–8, 109
Gruithuisen, P., 129, 178
Gylippus, 139

Haas, W. H., 125
Hadley Delta (lunar peak), 104
Rill, 170
Halley's Comet, 26
Haloes, lunar, 66
Hansen's theory, 150, 157
Harriott, T., 19–20, 75–6, 157
Hartman, T., 68
Harvest Moon, 52–4
Hatfield, Commander H. R., 93
Hermes, 24
Herschel, Sir J., 90, 129–32, 198
Sir W., 20, 83, 120, 128, 129
Hertzsprung (lunar crater), 184
Hevelius, 79–81, 139
Hoba West Meteorite, 27
Hooke, R., 190
Hörbiger, H., 174
Hot Spots, lunar, 141–2
Hraunbunga, 189
Hunter's Moon, 54
Huygens, C., 83
Hverfjall, 188–9
Hyginus N (lunar feature), 200, 248

I.A.U. lunar map, 91
Icarus, 24
Iceland, 110, 155

Ingenii, Mare, 157, 184, 312
Insects, lunar?, 133
Io, 24
Ionosphere, the, 63
Irwin, J., 104, 109, 170

Jackson-Gwilt (lunar crater), 259
Jodrell Bank, 148, 155
Jupiter, 24, 32, 126
  occultation of, 122, 142

Kaufman, L., 68
Kepler, J., 19, 48, 79, 127, 139, 144
Klein, H., 200, 205
Kohoutek's Comet, 26
Kopal, Z., 175, 177, 185, 187
Kordylewski, K., 42–3
Kordylewski's 'cloud', 42–3
Korolev (lunar crater), 184
Kozyrev, N. A., 204–5, 280
Krakatoa, 140
Kuiper, G., 92, 106, 179

L.R.V. (Lunar Roving Vehicle), 164, 169
Laplace, P. S. de, 31
Leibnitz (lunar crater), 184
  Mountains, 295
Leibowitz, H., 68
Leonardo da Vinci, 16, 79
Librations, lunar, 56–7
Liceti, F., 79
Lippershey, H., 74
Lipski, Y. N., 124
Locke, R., 130–1
Loewy, M., 90
Lohrmann, W., 85, 197–9
Lovell, J, 94, 113
  Sir B., 148, 155
Lower, Sir W., 76–7
Lucian, 144
Lúdent, 188
Lunabase, 101, 157
Lunar Base, 12, 20, 214–5
'Lunar City' (Gruithuisen), 129
Lunar Hoax, 129–32
Lunarite, 101
Lunation, the, 47
Luna-1, 93, 147–8
  -2, 93, 147–8
  -3, 13, 57, 93, 147, 149, 152–3
  -4, 154
  -9, 155
  -13, 156
  -16, 161, 285
  -20, 161
Lunokhod-1, 161
  -2, 161, 234, 248
Lyot, B., 124

Mädler, J. H., 85, 88, 129, 139, 197–9, 200, 202
Mädler's Square, 200–1
Magma, 164
Magnification for observations, 220
Man in the Moon, 14
Mare Æstatis, 268
  Cognitum, 284
  Veris, 268
Maria, lunar, 99–101
  ages of, 168
  list of, 311
  origin of, 105, 182
Mariner-9, 23, 40, 193–4
  -10, 23, 195
Mars, 22–3, 32, 34–5, 126, 165, 167, 193–4
Mascons, 158–60
Maskelyne, N., 128
McCall, G. J. H., 171
Mendeleev (lunar crater), 184, 312
Mercury, 21–3, 36, 54, 126, 165, 167, 195
Meteor Crater, 27, 110, 179–80, 181, 188–9
Meteorites, on Earth, 26–8
  on Moon, 171
Meteors, 26
  lunar, 124–5, 143
Middlehurst, B., 207–8
Mills, A., 192–3, 210
Minor Planets, 23–4
Mitchell, T., 169, 210
Miyamoto, S., 191, 208–9
  lunar atlas by, 93
Monte Somma, 191
Moon, age of, 32–6
  atmosphere of, 120–6
  changes on, structural?, 197–201
  concavity of orbit, 50–1
  core of, 35, 166
  data concerning, 227
  far side of, 13, 149–53, 157–8, 312–4
  future of, 39
  internal structure of, 165–8
  landings on, 230–1
  life on?, 126–34
  lithosphere of, 167
  lunacy, connection with?, 69
  magnetic field of, 33, 166–7
  maps of, 74–88, 91–5
  maps, naked-eye, 74
  movements of, 45–59
  myths, 14–16
  observing hints for, 217–21
  orange soil on, 169
  origin of, 31–7
  phases of, 45–6
  plants, effects on?, 70

pole star of, 215
recession of, 37–9
rotation, synchronous, 54–5, 191
satellite of?, 44
status of, 21, 36
travel to, 144–5
variation in apparent size of, 49
weather, connection with?, 65–6
worship, 14, 16
Moon Illusion, 67–9
Moonblink device, 207, 222–3
Moonquakes, 167, 210–11
Moscoviense, Mare, 153, 157, 184, 312
Mountains, lunar, 101
  heights of, 20, 78, 101, 104–5
  nature of, 102–3
Muller, P., 159

Nasmyth, J., 189–90
Nebular Hypothesis, 31
Neison, E., 86, 89, 198, 201
Neptune, 24
Nicias, 138–9
Nininger, H. H., 73, 200
Nodes, lunar, 51–2
Nomenclature, lunar, 81, 96–7, 158, 312

Ocampo, S., 175
Occultations, lunar, 120–3, 224
O'Keefe, J., 35
Olympus Mons, 191, 193
Öpik, E. J., 124, 179, 205
Orange soil, lunar, 169
Orbiter probes, 93–4, 156
  -5, 159
Orientation, lunar, 82–3, 93–4, 96, 232
Ovenden, M. W., 176

Paraselenæ, 66
Paris Atlas, lunar, 90
Peal, S. E., 174
Pelée, Mont, 140
Peloponnesian War, 138
Perseid Meteors, 26
Phobos, 22–3, 40, 54, 194
Photography, lunar, 90–3, 222
Pickering, W. H., 33–4, 44, 91, 122–3, 133, 142, 203
Planck (lunar crater), 312
Planets, 22
  earthquakes, connection with?, 65
  occultations of, 122
Plateaux, lunar, 117
Pleiades, occultation of, 121, 122
Plutarch, 18
Pioneer probes, 24
Pluto, 25
Proctor, R. A., 178

Ptolemy, 17–19, 48, 67–8
Puiseux, P., 90

Quarantining of astronauts, 134

Radon emissions from Moon, 207
Rainbows, lunar, 67
Ranger-7, 93, 154
  -8, 154
  -9, 114, 154, 280
Rays, lunar, 98, 117–18, 187–8
Regolith, lunar, 164–5
Regulus, occultation of, 121
Riccioli, G., 81, 263
Ridges, lunar, 107
Rigel, 29
Rills, lunar, 83, 108
Roche Limit, 39, 177
Rock, I., 68
Rockets, principle of, 145–6
Rocks, lunar, age of, 32, 168
  materials in, 164
Rømer, O., 28
Rosse, 4th Earl of, 173
Rutherfurd, L., 90
Ruud, I., 177

Saari, J. M., 141
Sacred Stone of Mecca, 27
Saros, the, 138
Sartory, P. K., 222
Satellites, origin of, 40
Saturn, 24–5
  occultation of, 123
Saunder, S. A., 91, 142
Schlüter (lunar crater), 272
Schmidt, J., 87, 88, 197–9
Schmitt, H., 125, 162, 169, 213
Schrödinger (lunar crater), 157, 312
Schröter, J. H., 83–5, 120, 128, 199, 201–2
Scott, D., 104, 109, 170
Secchi, A., 198
Secular Acceleration, 58
Seismometers, lunar, 167–8, 210–11
Selenographical Society, 89
Shadows, lunar, 97
Shaler, N. S., 151, 182
Shepard, A., 162, 163, 169, 210
Shorthill, R. W., 141
'Shorty' (lunar craterlet), 169
Siberian Meteorite, 27, 181
Sjögren, W. L., 159
Solar cycle, 210
Solar System, the, 21–8
  origin of, 31–2
Solar wind, 163
Spurr, J. E., 101, 103, 107–8, 117, 191
Sputnik 1, 12, 146
Stars, the, 28–9

Stickney (crater on Phobos), 194
Strout, E., 75
Sun, evolution of, 39–40
Surveyor probes, 155–6
Surveyor-3, 171
  -7, 118
Synodic month, the, 47

T.L.P. (Transient Lunar Phenomena),
  89, 202–12, 222–3
Tamrazyan, G. P., 64
Tektites, 70–3
Telescopes, 217–19
Terminator, lunar, 97–8
Thales, 17, 138
Thornton, F. H., 201
Tides, atmospheric, 63–4
  land, 63
  oceanic, 14, 60–4
Titan, 25, 36
Tombaugh, C., 42, 44
Toto, 41
Triton, 25, 36
Trojan asteroids, 24, 42–3
Trouvelot, E., 289
Tsiolkovskii, K. E., 145–6
Tsiolkovskii (lunar crater), 101, 153,
  157–8
Twilight, lunar, 202

Uranus, 24
Urey, H., 35, 37, 179, 185

V.2 weapons, 146
Valleys, lunar, 105
van de Graaff (lunar crater), 167
Velikovsky, I., 174, 176
Venus, 23, 36, 121, 167, 194–5, 202
Verbeek, R. D. M., 70, 72
Verne, Jules, 41–2, 145
Vesta, 24
Vesuvius, Mount, 191
Vladivostok Meteorite, 27, 181
von Braun, W., 146
Vredefort Ring, 27–8, 180

WEL, 174
Wargentin, P., 139
Warner, B., 205
Weekes, 63
Weisberger, Herr, 175
Wells, H. G., 129, 132
Wesley, W. H., 91
Whipple, J. A., 90
Wilkins, H. P., 91, 149, 151–2
Wilkins, J., 144
Wolf Creek Crater, 180
Wood, R. W., 261
Wood's Spot, 262
Wrinkle-ridges, lunar, 104, 107

Xenophanes, 17–18

Zeta Draconis, 215
Zodiacal Light, 28